D0151139

Scientific Teaching

SCIENTIFIC TEACHING

Jo Handelsman

Sarah Miller

Christine Pfund

THE WISCONSIN PROGRAM FOR SCIENTIFIC TEACHING
Supported by the Howard Hughes Medical Institute Professors Program

ROBERTS AND COMPANY
Englewood, Colorado

W.H. FREEMAN AND COMPANY
New York

Publisher:	Sara Tenney
Executive Editor:	Kate Ahr
Acquisitions Editor:	Marc Mazzoni
Marketing Manager:	Deb Clare
Cover and Text Designer:	Mark Ong, Side By Side Studios
Production Coordinator:	Ellen Cash
Composition:	Mark Ong
Printing and Binding:	RR Donnelley

On the cover: Fractal image "Meduses" by Jean Marc Silvestre.
Used with permission of the artist.

THE WISCONSIN PROGRAM FOR SCIENTIFIC TEACHING
Supported by the Howard Hughes Medical Institute Professors Program
Contact Information:
Phone: (608)265-0850
E-mail: scientificteaching@mailplus.wisc.edu
Supplementary materials are available online at http://scientificteaching.wisc.edu

Development of this book was supported by a grant from the Howard Hughes Medical Institute to the University of Wisconsin-Madison in support of Professor Jo Handelsman.

Library of Congress Control Number: 2006937149

ISBN-13: 978-1-4292-0188-9
ISBN-10: 1-4292-0188-6

Published in collaboration with Roberts & Company Publishers

Printed in the United States of America

Fifth printing

W. H. Freeman and Company
41 Madison Avenue
New York, NY 10010
Houndmills, Basingstoke
RG21 6XS, England

www.whfreeman.com

In memory of Denice Denton

1959–2006

Contents

Foreword

T**he idea of an institute that immersed scientists** in issues related to undergraduate teaching and learning in the life sciences was first proposed by the authors of the 2003 report from the National Research Council, *Bio2010: Transforming Undergraduate Education for Future Research Biologists* (National Research Council 2003). Later that year, former National Academy of Sciences President Bruce Alberts approached two prominent biologists who are also known for excellence in teaching—Jo Handelsman and Bill Wood—to develop such an institute. For three years, this institute has been assisting scientists of all levels of experience to further engage in undergraduate education in a manner similar to how they conduct their own scientific research (a.k.a. **scientific teaching**). Simultaneously, Handelsman's Wisconsin Program for Scientific Teaching has trained dozens of graduate students and postdocs to teach according to the tenets of this approach.

Guiding scientists in becoming more effective teachers is no small feat. Few college faculty in the natural sciences receive formal preparation for teaching or know much about the emerging research about how people learn. Faculty continually struggle with how to apportion the time and effort that they are expected to devote to their research, teaching, and service. Too often, teaching—especially in courses with large enrollments or targeted to non-science majors—becomes a secondary pursuit because it is not rewarded in the same ways or to the same degree as research. Thus, the need for a guide that can help faculty think more comprehensively about their teaching and, most importantly, their students' learning in ways that are based on emerging research, is vital and critically timely.

Scientific Teaching builds upon and greatly expands the message and resources from the paper with the same title that Handelsman, Wood, and their colleagues published in *Science* in April 2004. This guide to improving undergraduate education in the STEM disciplines integrates seamlessly the issues that most undergraduate faculty and teaching assistants confront over the course of their teaching careers and offers practical advice about ways to truly improve the undergraduate science experience. *Scientific Teaching* also helps instructors in higher education explore issues that are either difficult to discuss with colleagues or which may sometimes be overlooked: the nature, dimensions, and benefits of student diversity in a classroom; the improvements that accrue to both instructor and student when emphasis is shifted from teaching to student learning; what happens when assessment becomes both integral and integrated into every facet of a course in ways that help measure real learning.

Through numerous examples and references to the literature, Handelsman, Miller, and Pfund emphasize the principles and applications of scientific teaching. The book can serve as the basis for helping students to learn science, to gather data about whether and how much they have learned, and to make mid-course corrections based on those data. *Scientific Teaching* enables faculty in the natural sciences to apply to classroom settings what they do every day as researchers.

It is for all of these reasons that *Scientific Teaching* should become a fixture in the office of every natural sciences faculty member in higher education who wishes to think more deeply about and to improve her or his teaching.

Jay B. Labov
Senior Advisor for Education and Communications
The National Academies

Preface

Revisiting the One-Handed Pianist

Envision music schools that train pianists to play with their right hands, hoping the left hands will figure it out all on their own.* As unthinkable as this may seem, it is not unlike the way research universities train scientists. Just as piano playing is a two-handed job, so is the mission of higher education: to generate **and** disseminate knowledge. Why, then, do we take preparation for one part of this endeavor so seriously and treat the other so casually?

In the impressive assemblage of 4,000 colleges and universities in the U.S., the prerequisite for most teaching positions is a Ph.D.—a degree wrought of rigorous, prolonged training in the conduct of research. The Ph.D. is arguably the most successful element of the U.S. educational system. The results are clear—years of course work, exams, thesis committee meetings, conducting research, seminars, and painstaking dissertation writing and rewriting produce scientists with research skills honed to a fine point. Although graduate study affords excellent preparation for many careers because it teaches people to think, reason, and analyze, many Ph.D. recipients enter careers in teaching, for which they have no formal training. Most survive in these jobs, and some even excel, but how many more would excel if they were afforded training in teaching that approached the quality of their training in research? What would be the impact on the students they go on to teach?

The impetus for this book is to adjust the historic imbalance by offering current and future faculty and instructors streamlined, accessible training in teaching. Our goal is to make teaching more scientific; thus the title, *Scientific Teaching*. Embedded in this undertaking is the challenge to all scientists to bring to teaching the critical thinking, rigor, creativity, and spirit of experimentation that defines research.

* The one-handed pianist was first visited in Jo Handelsman's 2003 article, Teaching scientists to teach. *HHMI Bulletin* 12:31.

If this book is successful, our readers should attain a sense of mastery in teaching. When people achieve mastery of a new skill or field, or when they learn something that induces a fundamental or global change in their attitudes or understanding, they often want to share their discovery or mastery with others. This book is therefore intended for two types of educators, or perhaps more accurately, educators at two different points on the continuum of change. The first group is comprised of individuals who simply want to learn more about teaching and the scholarship about it. For those, the first five chapters will provide a short but in depth treatment of the subject. For those who have practiced scientific teaching and want to share it with others, Chapter 6, "Institutional Transformation," will provide the next steps. In Chapter 6, we provide ways to consider institutional change and then present a series of workshops that can be offered to graduate students or colleagues about each of the aspects of scientific teaching explored in the first five chapters. We expect that these sections may be important to educators at different times in their careers—when starting out, the "what" and "how to" sections may be most critical, but over time and with experience, the sections addressing "how to help others" may become more relevant.

Instructors often say that effecting change in their classrooms is limited by time and resources. We hope, therefore, to make the acquisition of this field of knowledge efficient. This book is intended to make it easy for instructors to learn about the theory and practice of teaching and the research that indicates what works. We hope that instructors will find *Scientific Teaching* a one-stop shop that introduces them to specific strategies to change their classrooms and their campuses and find resources that will provide them with other approaches and examples. The book is intended to guide new instructors to develop sound teaching habits based on evidence rather than experience alone and to aid veteran instructors in enriching and reinvigorating their teaching with proven methods.

None of the methods presented in *Scientific Teaching* are entirely new. Rather, we have selected highlights of the literature, along with other ideas and techniques, and distilled them for instructors who don't want to research this vast territory themselves. We hope that the synthesis is new and that its application to science education is fresh, but we don't claim that many of the specifics originated with us. We have tried to provide citations to the original literature from which we have drawn, so that readers can access that literature. Our synthesis is by no means comprehensive. Entire schools of thought, experts, and methods have been ignored by necessity. We ask the indulgence of those whose favorite example is omitted but hope we have included the concepts that most educators consider to be central. Similarly, where we present controversial ideas (especially in Chapter 4, in which we are deliberately provocative) we hope our readers will appreciate the challenge to think critically about their teaching and their students' learning.

Science education has uncertain connections to education and an ambivalent relationship with science itself because of the apparent conflict between teaching and research responsibilities in many places of higher education. We intend this book to recast the relationships of science and education and between research and teaching. Science and education can cooperate to reform science education. Teaching can be a complement to research rather than being in conflict with it. Teaching is, after all, the mission that connects all members of academia, and scientific teaching draws on the tenets that are the very core of research. The fractal image on the cover of this book captures our goals. We hope to help untangle the current octopoid environment that so unnecessarily complicates science education, fostering more direct and positive linkages between science **and** education. And we hope that one day we will all become two-handed pianists.

JH
SM
CP
October, 2006

Acknowledgments

Just as teaching is a collaborative process between teachers and learners, the writing of this book has been a collaborative process involving colleagues, students, friends, family, and critics. But in this case, unlike the classroom, it has been harder to tell the teachers from the learners. Many people have taught us about teaching and a number of institutions have made possible that process, the emergence of scientific teaching, and this book itself. Although we can't thank everyone one of them individually, we send a collective "thank you!" to all of our HHMI Teaching Fellows, students in *Teaching Biology*, participants in the National Academies Summer Institute on Undergraduate Education in Biology, and many other students and colleagues for helping us learn how to teach. We would like to offer a special thanks to a few people:

Janet Branchaw, Janet Batzli, Michelle Harris, and Amy Moser and the many other colleagues who provided reviews of our students' instructional materials, and who stood aside for a moment to let the next generation of faculty begin to teach;

Whitney Robertson and Karen Cloud-Hansen, for their creative teachable units;

Kerry Brenner, Amy Chang, Bob DeHaan, Diane Ebert-May, Adam Fagan, Jim Gentile, Ishrat Khan, Jay Labov, Millard Susman, Lillian Tong, Bill Wood, and Bob Yuan for creating the Summer Institute;

Christine Maidl Pribbenow for educating us about assessment and qualitative research and separating the world into the possible and the impossible;

Helaine Kriegel and Barbara Houser for bringing biology to life;

Jim Stewart for teaching us about backward design and much more;

Christina Matta, Alison Duff, and Jennifer Sheridan for research, writing, and editing;

Bob Mathieu, Aaron Brower, and the Delta Program in Research, Teaching, and Learning for collaborating in the reform of undergraduate science education;

Alice Pawley for her insights on bringing diversity into practice;

Sandra Courter for developing and sharing the idea of reading assessments;

Nancy DiIulio, Clarissa Dirks, Michael Hanna, Robin Heyden, Jay Labov, Diane O'Dowd, Marilla Svinicki, and Brad Williamson for their astute reviews of the manuscript;

Hilary Handelsman for many rounds of meticulous editing and her appreciation of Horton's appearance in Chapter 2;

Mark Salzwedel for more rounds of meticulous editing;

Mark Ong at Side by Side Studios for composition and design;

John Hastie for graphic design;

Jean-Marc Silvestre for creating and contributing the beautiful cover image;

Bill Wood for his friendship and for embodying scientific teaching;

Paul Williams for setting the standard to which we all aspire;

Peter Bruns for being the creative spark that makes so much happen;

Lillian Tong for her insight into instructional material design;

Chin Sun for helping us construct the meaning of learning;

Ben Roberts for his wonderful insights, suggestions, and unflagging enthusiasm for scientific teaching;

Sandra Gossens for keeping *everything* running properly;

Bruce Alberts and the National Academy of Sciences for advocacy for science education and the Summer Institute;

Tom Cech, Peter Bruns, Maria Koszalka, Mary Bonds, Jennifer Donovan, Steve Barkanic, Emily Brownold, and many others of the Howard Hughes Medical Institute who have made the Professors Program such a success;

The University of Wisconsin-Madison for its steadfast support of education and innovation;

And finally, we, and the entire community of science educators, owe a great debt to the **Howard Hughes Medical Institute** for its vision and leadership in the unification of science and education.

> I am most proud of seeing the pattern where there appears not to be any, for example, as when rose petals fall to the ground. I separate them or break them down into two main parts, and then build a symmetrical arrangement.
>
> —David Ausubel, 2006

CHAPTER 1

Scientific Teaching

Scientific Teaching Defined

The goal of scientific teaching is to make teaching more scientific. Embedded in this undertaking is the challenge to all scientists to bring to teaching the critical thinking, rigor, creativity, and spirit of experimentation that defines research. Scientific teaching also posits that the teaching of science should be faithful to the true nature of science by capturing the process of discovery in the classroom (Handelsman *et al.* 2004). There is evidence that teaching scientifically improves undergraduate education and student learning, and this evidence needs to inform instructional decisions. We have tried in this book to provide an overview of the complex and expansive landscape of evidence, approaches, methods, and theories. However, teachers of science must customize the methods for their own teaching styles, curricula, goals, and institutions.

Why Scientific Teaching?

Good science

Scientific teaching is needed because science is important. We do a disservice to our discipline and our students by reducing science education to a spontaneous, sometimes haphazard, process of delivering information with no attention to evidence either from the published literature or from our students about the validity of our delivery methods. Scientific teaching employs methods whose effectiveness has been established by research, it promotes student assessment of their own learning, and it depends on mid-course corrections in response to formal and informal assessments of learning. We simply cannot afford to discover only when we grade exams, long after the actual teaching event, that our students largely missed or misunderstood the content and concepts. The rapid expansion of many scientific frontiers places the onus on science educators to teach efficiently and effectively, assuring that students acquire a vast amount of knowledge and retain a good portion of it. Our students, whether they major in biology, art history, math, or elementary education, should not complete their college education without understanding basic principles and facts about the world around them. Equally important, our students need to emerge with an understanding of the nature of science so they can appreciate the origins of scientific information, think critically about new problems and situations, and sustain a lifelong curiosity about the world around them.

Good science education

Most science is taught as lectures that are dominated by facts rather than principles and ways of thinking. Sadly, as will be discussed later in this chapter, substantial evidence shows that lecturing alone is a relatively ineffective way of teaching, and retention from lectures is poor. In 2003, the National Research Council published *Bio2010* (National Research Council 2003), a report that offered recommendations for new biology curricula that better reflect the nature of science as a dynamic, interdisciplinary, and rapidly changing frontier. *Bio2010* echoes the 1989 call for a scientifically literate public by the American Association for the Advancement of Science in *Science for All Americans* (Rutherford 1990), which charged higher educators with teaching scientific habits of mind. In addition, many employers are demanding more efficient and effective employees with problem-solving capabilities, the ability to work in teams, and analytical skills, yet many of our students are unprepared for these demands.

The traditional college curriculum, designed around information and factual knowledge, provides insufficient education for the new scientific workforce. The national reports recommend that college courses retool their goals to highlight conceptual understanding, interdisciplinary context, authentic scientific experience, and interpersonal skills. Transforming classrooms to emulate the true human, dynamic enterprise of science is likely to increase the heterogeneity of students who are attracted to scientific fields of study, and it will augment all students' marketable skills. In short, practicing real science in the college classroom benefits everyone.

Excerpt from *Science for All Americans* (Rutherford, 1989)

"Education has no higher purpose than preparing people to lead personally fulfilling and responsible lives. For its part, science education—meaning education in science, mathematics, and technology—should help students to develop the understandings and habits of mind they need to become compassionate human beings able to think for themselves and to face life head on. It should equip them also to participate thoughtfully with fellow citizens in building and protecting a society that is open, decent, and vital. . . .

"Science, mathematics, and technology do not create curiosity. They accept it, foster it, incorporate it, reward it, and discipline it—and so does good science teaching. . . .

"Science, mathematics, and engineering prosper because of the institutionalized skepticism of their practitioners. Their central tenet is that one's evidence, logic, and claims will be questioned, and one's experiments will be subjected to replication. In science classrooms, it should be the normal practice for teachers to raise such questions as: How do we know? What is the evidence? What is the argument that interprets the evidence? Are there alternative explanations or other ways of solving the problem that could be better? The aim should be to get students into the habit of posing such questions and framing answers."

Good teaching

Alternatives to lecturing abound, but most college and university instructors lack the background to design or select teaching methods wisely. Scientific teaching aims to bring a philosophy and framework to teaching that makes the process more rigorous, reflective, and evaluative. Scientific teaching is part of a growing effort to improve teaching and learning at the post-secondary level. Other efforts are described by such terms as *classroom assessment* (Cross and Angelo 1988, Angelo and Cross 1993, Angelo 1998), *classroom research* (Cross

1990, Cross and Steadman 1996), *the scholarship of teaching and learning* (Glassick *et al.* 1997, Huber 1999 and 2005, Hutchings and Shulman 1999, Hutchings *et al.* 2002), *action learning* (Kember 2000), *classroom action research* (Mettetal 2001), *research-led teaching* (Brew 2001), and *teaching-as-research* (CIRTL 2006; Connolly *et al.* in press). This book is designed to demystify the process of scientific teaching and bring it to the forefront of intellectual approaches among the professoriate.

Models for Learning

Any discussion about scientific teaching necessitates a discussion about learning. There are many ways to formulate the learning process, and there are many theories about learning. Different disciplines approach learning with unique questions and ways to test the validity of the theories. In addition, the term *learning* is used differently in different contexts. Some argue that learning is the same as knowing, others equate learning with understanding, and still others argue that true learning is apparent only if students can apply their knowledge and understanding. While all these positions have merit, for the purposes of this book, **we use the term learning to indicate the acquisition of knowledge and understanding, the development of skills, and the ensuing changes in affect or behavior.** For example, when we describe learning goals, we follow up by asking: What will students know, understand, and be able to do?

The information provided here is by no means comprehensive, but it provides an overview of some of the most influential theories and models of learning. Some of the models are controversial, and some have been more rigorously evaluated than others, but all of them have merit and are worth considering in scientific teaching. What researchers do agree upon is that successful learners are actively engaged, they receive regular feedback, and they are able to integrate and scaffold new information in new contexts. The purpose in providing the following formulations is to give some concrete structure to an abstract process, which can serve as an organizing tool for teachers. Emerging research from neuroscience, cognitive psychology, and education—as well as research stemming directly from undergraduate science classrooms—will continue to shape this dynamic and complex field. Later in this chapter, we will discuss learning outcomes and ways to measure these aspects of learning.

"How People Learn"

It is essential to teach in a way that facilitates the kind of learning desired. In a compendium of resources entitled, *How People Learn*, the National Research Council (NRC) (National Research Council 1999a) provides an overview of perspectives about learning from multiple disciplines.

According to the NRC report, different learning experiences uniquely shape each person's ability to interpret and acquire knowledge. Many factors affect the receptiveness of the brain to learning, including what the student brings to the classroom, the instruction, and the learning environment. Students bring to the classroom not just their prior knowledge (which may or may not be accurate); they also bring past experiences, culture, gender, race, ethnicity, beliefs, sexual orientation, physical condition, and psychological status. In addition, different teaching methods reach a variety of students in various disciplines and topics. The adept teacher is sensitive to which concepts in each discipline are easy or difficult for most students to grasp, and adjusts instruction accordingly so that it drives learning in a direction that parallels the nature of science. The classroom environment also affects learning. Specifically, the physical environment, the human community, and the attitudes toward learning contribute to or detract from that process. One of the most compelling findings is a correlation between how students study and physical changes in their brains. Specifically, the intensity and duration of experience in a complex environment corresponds with the degree of structural change in the brain. The structural changes can reorganize the functions of the brain and lead to subsequent changes in behavior (National Research Council 1999a). Not surprisingly, less stimulating experiences yield fewer changes in the brain than experiences in a more complex environment.

The NRC has also published a companion report, *How People Learn: Bridging Research and Practice* (National Research Council 1999b). This report consolidates key findings about learning and provides recommendations for teaching and classroom environments. (See excerpt, next page.)

Constructivism

Perhaps the most widely accepted and useful ideas about learning are derived from **constructivism**, a robust theory about learning that proposes people learn by constructing their own knowledge. Constructivism emerged from the work of a series of great thinkers led by John Dewey and David Ausubel who have

Excerpt from *How People Learn: Bridging Research and Practice*

(National Research Council 1999b)

Key findings about learning:

1. Students come to the classroom with preconceptions about how the world works. If their initial understanding is not engaged, they may fail to grasp the new concepts and information that are taught, or they may learn them for purposes of a test but revert to their preconceptions outside the classroom.
2. To develop competence in an area of inquiry, students must: (a) have a deep foundation of factual knowledge, (b) understand facts and ideas in the context of a conceptual framework, and (c) organize knowledge in ways that facilitate retrieval and application.
3. A "metacognitive" approach to instruction can help students learn to take control of their own learning by defining learning goals and monitoring their progress in achieving them.

Implications for teaching:

1. Teachers must draw out and work with the preexisting understandings that their students bring with them.
2. Teachers must teach some subject matter in depth, providing many examples in which the same concept is at work and providing a firm foundation of factual knowledge.
3. The teaching of metacognitive skills should be integrated into the curriculum in a variety of subject areas.

Designing classroom environments:

1. Schools and classrooms must be learner centered.
2. To provide a knowledge-centered classroom environment, attention must be given to what is taught (information, subject matter), why it is taught (understanding), and what competence or mastery looks like.
3. Formative assessments—ongoing assessments designed to make students' thinking visible to both teachers and students—are essential. They permit the teacher to grasp the students' preconceptions, understand where the students are in the "developmental corridor" from informal to formal thinking, and design instruction accordingly. In the assessment-centered classroom environment, formative assessments help both teachers and students monitor progress.
4. Learning is influenced in fundamental ways by the context in which it takes place. A community-centered approach requires the development of norms for the classroom and school, as well as connections to the outside world, that support core learning values.

become the lodestars of modern education theorists. Early in the twentieth century when American students were schooled in a system based on authoritarian teaching and rote learning, John Dewey asserted the role of the learner and the learner's prior knowledge in the learning process, introducing his now-famous metaphor that students are not simply "empty vessels" to be filled by teachers (Dewey 1916). Dewey provided the blueprint for modern constructivist theory, formulated by David Ausubel, which contends that learning must accommodate and build upon the experience of the learner, who actively integrates new knowledge into a framework. Constructivism asserts that each of us creates our own rules and explanations, which we use to make sense of our experiences. Learning, therefore, is the process of adjusting our mental models to accommodate new experiences. Students assimilate new information by constructing scaffolds on which they can organize and retain facts and concepts (Ausubel 1963 and 2000). Teaching, in turn, must accommodate—indeed, augment—this process. Ausubel believed that instructors could provide the scaffold to organize knowledge, but that students needed to be intellectually engaged for the scaffold to be used to accomplish learning. Modern constructivists maintain that active participation helps students learn as well as develop the habits of mind that drive science.

Motivation, metacognition, thinking, and knowledge

In his 1998 report, *Theory-Based Meta-analysis of Research on Instruction*, Robert Marzano presented a synthesis of learning models based on research about how people process information. Marzano contends that, before any learning happens, students need to "cross the Rubicon." By this, he means that they must decide to engage in learning. Once the Rubicon is crossed, three steps need to happen, from a cognitive perspective, for learning to take place. First, the brain determines what goals will be achieved, what strategies will be employed, and what methods will be used. This process of **metacognition** keeps the learning process moving forward and continually interacts with the next step, **cognition.** Cognition involves thinking strategies that process and analyze information and allow the brain to determine how it is used. Finally, the brain acquires **knowledge**, which can include information, mental processes, or skills. If the brain disengages at any point before then, learning simply won't happen (Marzano 1998).

Too often, grades are the motivation for learning. While this is certainly one way to engage students, the outcome is that students are motivated to get to the end point (learning) as quickly and efficiently as possible, which can occur at the expense of deep learning. Marzano (1998) proposed instructional strategies that build on theories about human information processing, some of which are used as the foundation for examples in Chapters 2 and 3 of this book.

Learning styles

Different people learn in different ways and even the same person may learn a variety of things using a range of strategies. There is no question that diverse strategies exist for acquiring new information, understanding, or skills, but there is no single organizing theory or body of research that centers the literature on learning styles. Therefore, the following section is intended to familiarize readers with the language of learning styles and summarize the findings about their use in teaching. Studies of learning styles do not offer definitive instructions to teachers, but the simple recognition that many different learning styles may be at work in a single classroom highlights the teacher's challenge and guides the development of instructional strategies.

The research on learning styles is convoluted and at times contradictory. Perhaps the best characterization is that there is a continuum of ways to approach learning styles. In an extensive review of more than 3800 publications, a team of researchers at the Learning and Skills Research Center in London constructed a model to organize the theories (Coffield *et al.* 2004). They classified learning style models into five "families" (Table 1.1).

At one end of the landscape, researchers suggest that learning styles and preferences are biologically based and therefore fixed. A widely used example of this family of learning style theories is the "VAK" model, in which learners are classified as visual, auditory, or kinesthetic according to results of a learning styles inventory. Learning styles, the proponents of this type of model contend, should be accommodated in the classroom. The three families in the middle of the continuum vary in the degree of "stability" they attribute to learning styles. These families are based on theories that range from cognitive structure (more stable) to personality type to learning preference (more flexible).

At the other end of the spectrum is a family of models based on the idea that learning styles are only one aspect of teaching and learning. They distinguish between learning styles and other facets of learning—for example, strategies that students use in learning—which the developers consider to be more significant

Table 1.1 The landscape of learning styles.

Coffield *et al.* (2004) identified over 70 unique approaches to learning styles, which they classified into five overarching "families." The families range from models that explain learning styles as innate (and therefore stable) on the left, to models that regard learning styles as "flexibly stable" or as a single component among many factors that contribute to learning efficacy. (Adapted from Coffield *et al.* 2004.)

Learning styles and preferences are innate.	Learning styles reflect deep-seated features of the cognitive structure.	Learning styles are one component of a relatively stable personality type.	Learning styles are flexibly stable learning preferences.	Learning styles are only one component of learning and not necessarily fixed.

in shaping learning. Advocates of these models contend that all of these characteristics need to be factored into instructional design and that the developers consider learning styles to be flexible attributes that may change over time or with experience. For example, factors such as motivation, social dynamics, and course culture affect how a student chooses to approach—or avoid—learning, and therefore should be considered at least as much as learning styles.

Coffield's team analyzed representative models from each family for their reliability and validity. The primary findings from the report are that:

1. Learning style models can be used as a tool to encourage student self-awareness and improve learning skills by diagnosing how they learn and by showing them how to enhance their learning. To do this, however, instructors need to develop reliable and valid instruments to measure the learning styles in question. In the literature, no single learning style model has proven sufficiently effective to serve as the basis for wide-scale curricular reform based on examination for reliability and validity.
2. Learning style models can provide learners with a "lexicon of learning." The language can be used as the basis for students to discuss their own learning preferences and those of others, to consider how people learn or fail to learn, and to account for how instructors can facilitate or hinder these processes.
3. Learning styles are only one way to consider learning in the classroom. Instructors should also consider the findings by the two of the largest

meta-reviews of educational interventions: Hattie (1999) and Marzano (1998). Together, these reports indicate that:

a. The most effective methods for improving student learning are reinforcement of the material and providing regular feedback to students about learning.
b. Instruction that takes a scientific teaching-type approach (*i.e.*, instruction that requires a metacognitive approach of setting learning goals, choosing appropriate strategies, and monitoring progress) are more effective than those which simply engage students.

Two of the most promising learning style models and inventories for higher education situations are the **Approaches and Study Skills Inventory for Students** (ASSIST, Entwistle 1990) and the **Inventory of Learning Styles** (ILS, Vermunt 1996). ASSIST is a multi-faceted inventory of learning that differentiates between "deep" and "surface" learning. The premise of ASSIST is to connect learning behaviors (motivation, study methods, and academic performance) with instruction (course design, environment, and assessment methods) in meaningful ways that guide instructional decisions. According to Coffield, the ASSIST model "is useful as a sound basis for discussing effective and ineffective strategies for learning and for diagnosing students' existing approaches, orientations, and strategies. It is an important aid for course, curriculum, and assessment design. . . . It is crucial, however, that the model is not divorced from the inventory . . . and that students are not labeled as 'deep' or 'surface' learners."

Similarly, ILS integrates multiple types of experiences into consideration of learning styles. Factors considered in this model include (1) how students process information, (2) why they choose to process it that way, (3) how they feel about the processes, (4) how they perceive learning, and (5) how students plan and monitor learning. ILS recommends a metacognitive approach to learning that involves self-directed assessment and regulation of behaviors. Coffield states, "ILS can be safely used in higher education, both to assess approaches to learning reliably and validly, and to discuss with students changes in learning and teaching."

Learning goals and intended outcomes

Understanding how students learn is important, but it is just as critical to set learning goal priorities and determine how to define and measure learning.

Knowing which learning goals are most important will help students focus on **understanding**. According to Wiggins and McTighe (1998):

> Understanding involves sophisticated insights and abilities, reflected in varied performances and contexts. We also suggest that different kinds of understanding exist, that knowledge and skill do not automatically lead to understanding, that misunderstanding is a bigger problem than we realize, and that assessment of understanding therefore requires evidence that cannot be gained from traditional testing alone.

Learning goals are useful as broad constructs, but without defined learning outcomes, goals can seem unattainable and untestable. The outcome is to the goal what the prediction is to the hypothesis: Just as experimental outcome can verify or refute a hypothesis, a learning outcome provides a measure of attaining a learning goal. Therefore, learning goals and outcomes used in unison are a powerful combination to articulate where students need to be and how far they are toward reaching the goals you have set.

Students who understand a concept should be able to demonstrate what they know in multiple ways. For example, a student who understands photosynthesis might be expected to explain the role of chlorophyll, interpret data about the uptake rate of carbon dioxide, or compare and contrast gas exchange in aquatic plants and desert succulents. The six facets of understanding are (Wiggins and McTighe 1998):

1. Explanation
2. Interpretation
3. Application
4. Perspective
5. Empathy
6. Self-knowledge

In addition, others have provided lexicons for elucidating student actions that demonstrate understanding or abilities (Culver and Hackos 1982, Hogsett 1992, Bloom and Krathwohl 1956). Bloom's Taxonomy, for example, provides a simple hierarchy of intellectual behaviors that are important to learning, categorized into six levels of cognition.

Within the hierarchy, Bloom identifies learning behaviors that typify each level. The following list is one example of words used to describe the actions students are expected to take to demonstrate their knowledge, skills, and understanding.

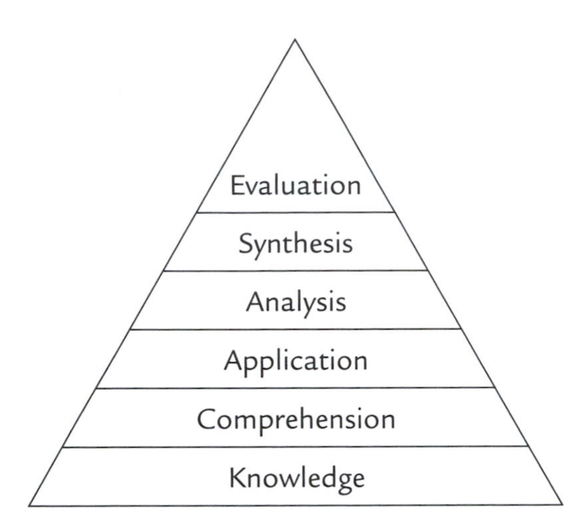

Figure 1.1 Bloom's Taxonomy. (Adapted from Bloom and Krathwohl 1956)

1. ***Knowledge***: arrange, define, duplicate, label, list, memorize, name, order, recognize, relate, recall, repeat, reproduce, state.
2. ***Comprehension***: classify, describe, discuss, explain, express, identify, indicate, locate, recognize, report, restate, review, select, translate.
3. ***Application***: apply, choose, demonstrate, dramatize, employ, illustrate, interpret, operate, practice, schedule, sketch, solve, use, write.
4. ***Analysis***: analyze, appraise, calculate, categorize, compare, contrast, criticize, differentiate, discriminate, distinguish, examine, experiment, question, test.
5. ***Synthesis***: arrange, assemble, collect, compose, construct, create, design, develop, formulate, manage, organize, plan, prepare, propose, set up, write.
6. ***Evaluation***: appraise, argue, assess, attach, choose, compare, defend, estimate, evaluate, judge, predict, rate, select, support, value.

Learning, in sum

In a 2005 review of evidence about undergraduate student learning, Robert DeHaan (2005) integrates findings from these and many other reports about learning and how they apply to instruction:

> Virtually all these expert sources agree that science education should focus: less on what instructors "cover," and more on what students learn and how well they

can use their knowledge; less on vocabulary and facts that students memorize, and more on students' understanding of scientific concepts and how those concepts fit together in a framework of knowledge about a subject; less on what students can repeat back immediately in class, and more on their long-term retention and ability to transfer their knowledge to contexts outside the classroom.

In essence, the most successful students are **curious**—interested in asking questions, investigating answers, and sustaining their own learning. Curiosity may, to a large extent, be innate, but it can also be learned. Classrooms that value independence, high standards, and conceptual knowledge help students become mindful of themselves as learners and thinkers. In this environment, students learn to confront misinformation, challenge previously held ideas, and construct new knowledge. Therefore, the most effective classrooms foster curiosity by engaging the diversity of students' minds through instructional design and classroom environment.

A Framework for Scientific Teaching

Teachers do not need to be experts in learning theory to be great teachers, but it helps to use methods that are (1) based on solid theory, (2) tested and evaluated in the classroom, and (3) organized by a logical framework. The methods presented in this book are based on constructivist theory. Some version of each of the methods has been studied and found to enhance learning compared with didactic approaches that involve the student as a passive partner. The logical framework is scientific teaching. (See Table 1.2.)

Scientific teaching offers a natural structure for teaching that parallels our approach to research. Rather than allowing a last-minute, somewhat-random approach to designing classes, building a framework according to the principles of scientific teaching requires forethought, planning, and time. As researchers, we would never start designing an experiment without first identifying a question, hypothesis, or outcome. A researcher would never sit down and say, "First, I think I'll run a gel. Then, maybe I'll take a look at my sample under the light microscope. Maybe at the end of the day, I'll do some immunoprecipitations. Tomorrow I'll develop a hypothesis and decide what I want to find out about my organism." Sound silly? In effect, this is what many scientists do in their teaching.

Scientific teaching requires that teaching be approached with an unremitting focus on outcomes. Instructors need to consider what they want their students to know, understand, and be able to do and work back from there. The good news is that once a sound framework is in place, filling in the details for each class session comes easily.

Backward design

An ideal structure, proposed by Wiggins and McTighe (1998), is known as **backward design.** Backward design involves three steps:

- Identify desired results (learning goals).
- Determine evidence for learning (learning outcomes and assessment).
- Plan learning experiences and instruction (activities).

Backward design is deceptively simple. It is familiar to scientists, because it is the process by which most experiments are planned, but it is the reverse of how most people conceive of their teaching. In backward design, the instructor articulates what students will learn *before* designing activities. Attention is thereby focused on successful learning, rather than exclusively on teaching. The learning goals serve as the foundation for the instructional materials, defining broadly what students will know, understand, and be able to do. Learning outcomes, in turn, describe what specific performances or behavior will indicate whether students have achieved the goals. Outcomes are measurable and may include detailed criteria that describe the different levels of performance. The criteria make clear how students will be evaluated and ensure regular feedback to the students about their progress. Once these pieces are in place, the learning experiences and assessments can be fleshed out. Together, the learning experiences, the assessments, and the written and visual materials used in teaching comprise the "instructional materials."

Beyond backward design: The role of review and revision in scientific teaching

Of necessity, a focus on outcomes creates a need for evaluation. Creating instructional materials is an important step, but to bring the scientific teaching process full circle, the instructor needs to teach the materials and determine whether the goals were met. The information gained from assessment will guide review and revision of the materials and may lead to new learning goals and instructional ideas.

Key questions in scientific teaching

- **Scientific content:** Do the learning goals represent the nature of science? (In other words, do students learn science together by questioning, investigating, analyzing, and discovering?)
- **Active learning, assessment, and diversity:** Do the activities engage a diversity of students in learning? Do the assessments measure progress toward the goals and provide useful feedback to both students and instructors?
- **Evaluation, review, and revision:** Were the goals achieved? What changes could improve instruction and learning?

Scientific Teaching in Practice

The following chapters provide an overview of research for scientific teaching topics and offer practical solutions for incorporating them into teaching.

Active learning and assessment

Chapters 2 and 3 focus on practical ways to use the concept of constructivism in the science classroom. These chapters explore the integration of active learning and assessment in scientific teaching and how to use these methods to help students become responsible learners. The ideas and tools presented are intended to help students learn to think like scientists: to acquire critical thinking skills, understand how to analyze information, provide justification for a solution to a problem, and continuously evaluate their learning. In addition, Chapters 2 and 3 provide an overview of research about active learning and assessment, and they highlight how active learning and assessment converge in practice.

Diversity

Chapter 4 explores the role of human diversity in the scientific classroom. Diversity is a lens through which instructors can evaluate instruction and learn-

Table 1.2 A framework for scientific teaching.

	Students Think Like Scientists	Teachers Teach Like Scientists
Goals and Outcomes	Students experience the nature of science, including discovery, inquiry, and experimentation. Students learn the skills and knowledge relevant to the discipline.	Teacher sets learning goals that represent the nature of science.
Assessment and Evaluation	Students learn to think critically and gauge their own learning; they know what criteria indicate successful learning.	Teacher sets clear criteria that establish the intended learning outcomes and determine how to measure learning.
Activities	Students engage in experiences that help them achieve the learning goals.	Teacher designs activities to help students achieve the learning goals.
Alignment	Students learn how to use feedback to monitor learning and adjust behavior accordingly.	Teacher determines whether the activities help students achieve the learning goals and whether the assessments effectively measure achievement. Teacher uses feedback from assessments to revise instruction.
Diversity and Collaboration	Students learn to work in groups and learn how to include a diversity of students in the process.	Teacher uses a variety of teaching methods to engage a diversity of students. Teacher discusses teaching methods, instructional materials, student diversity, and evaluation results with colleagues.

ing. Unfortunately, it's too often the part of teaching that "doesn't fit into the flow," that gets ignored in the interest of time, or that is simply too much trouble. Many instructors dismiss the importance of diversity with phrases like, "I treat all of my students the same." Others refuse to change teaching methods that present barriers to some students' learning, or fail to accommodate different students' needs. If instructors aim to help all students learn, then they need to consider the differences that exist among their students. Diversity matters, and it matters in ways unconscious, physiological, intellectual, and social. It affects how students behave, it affects their motivation, and it affects what they learn and how well they learn it. And quite significant, but unknown to many teachers, is the large body of research demonstrating that diversity is correlated with creativity and the success of teams and groups as well as the intellectual development of all students. Chapter 4 provides research, information, and suggestions for creating classrooms that are inclusive and respectful learning environments and engage a diversity of learners.

Building a teachable unit

Chapter 5 puts scientific teaching principles to work. It presents a framework for building a "teachable unit," which is an approach to developing instructional materials with goals and evaluation in mind. In this chapter, active learning, assessment, and diversity are used together in the development and review of instructional materials.

Institutional transformation

Chapter 6 offers insights into the nature of institutional change and a series of workshops that can be used to train a new generation of graduate students and postdocs to think scientifically about teaching before they enter the professoriate, or to build a community of colleagues that focuses on teaching science. The workshop series is intended to make it easy for instructors to disseminate the use of scientific teaching methods beyond their own classroom.

Every instructor can contribute to the transformation of undergraduate science education in many ways. Each of us can change our own teaching methods to enhance student learning as a first step. But science education needs and deserves a more comprehensive overhaul. There are many ways to effect global change. For example, discussions with colleagues—about research on student learning or to get feedback about a new set of instructional materials—can spread knowledge of scientific teaching and its outcomes. But is it possible for one person to make a difference in science education for a whole campus? The answer is a definitive *yes*. Chapter 6 presents strategies to instigate institutional change in science education by engaging administrators, training colleagues, and empowering the scientific teachers of tomorrow. Armed with knowledge, evidence, and activities, every one of us can transform science education locally or even globally.

The Lexicon of Scientific Teaching

Active learning
A process in which students are actively engaged in learning. It may include inquiry-based learning, cooperative learning, or student-centered learning.

Active learning exercise
An activity that actively engages students.

Alignment
Ensuring that the activities and assessments are designed to help students meet learning goals.

Assessment
Tools for measuring progress toward and achievement of the learning goals.

Backward design
Designing instructional materials by first setting learning goals, then determining what outcomes would illustrate achievement of those goals, and then designing classroom activities so that students meet the goals.

Biology concept framework
A model that provides a hierarchy of biology concepts and content for an introductory biology course. (Similar models also exist in chemistry and physics.)

Bloom's taxonomy
A hierarchy of intellectual behaviors that are important to learning, categorized into six levels of cognition: knowledge, comprehension, application, analysis, synthesis, and evaluation.

Cognition
The mechanisms the brain uses to process knowledge and analyze information.

Constructivism
The theory developed by education philosopher David Ausubel proposing that people learn by constructing knowledge.

Cooperative learning
The process by which students work together to solve a common problem.

Diversity
The breadth of differences that make each student unique, each cohort of students unique, and each teaching experience unique. Diversity includes everything in the classroom: the students, the instructors, the content, the teaching methods, and the context.

EnGaugement
An activity that simultaneously *engages* students in learning and *gauges* their understanding.

Evaluation
The process of analyzing the results of assessment and determining whether the goals have been achieved.

Group learning
The process of students working in groups to solve a problem. (See also: *cooperative learning.*)

Inquiry-based learning
The process of engaging students in the process of exploration and asking and answering scientific questions to acquire new knowledge and skills.

Institutional transformation
The process of changing the culture and practices of a campus to reflect a commitment to key ideals.

Instructional materials
The sum of materials and information—documents, notes, Web-based tools, activities, and handouts—that an instructor uses for teaching.

Knowledge
The information, mental processes, and skills that students gain by learning.

Learning
The acquisition of knowledge and understanding, the development of skills, and the ensuing changes in affect or behavior.

Learning goals
What students will know, understand, and be able to do.

Learning outcomes
Specific, measureable learning goals.

Learning styles
Preferences, approaches, and skills in learning.

Metacognition
The process of setting challenging goals, identifying strategies to meet them, and monitoring progress toward them.

Misconception
Incorrect understanding of a concept or fact.

Nature of science
Representation of science as a process that includes analysis, collaboration, communication, experimentation, evaluation, inquiry, and knowledge.

Prior knowledge
Collective knowledge, skills, and worldviews that students bring to a class.

Reading assessment
An exercise in which students develop an activity to gauge whether their peers understand a reading assignment.

Rubric
A tool that provides explicit criteria by which the work (homework, exam, project, etc.) will be judged.

Scientific teaching
Teaching science in a way that (1) represents the nature of science as a dynamic, investigative process based on evidence, (2) engages a diversity of people in a collaborative process and (3) has clear learning goals in mind, uses methods and instructional materials designed to improve student learning, and evaluates the methods iteratively.

Teachable tidbit
Part of a teachable unit. Example: a brainstorming activity. (See also: *active learning exercise* and *enGaugement.*)

Teachable unit
Instructional materials with clear learning goals that are designed to engage students in learning, provide feedback to both students and instructors about learning, and provide other instructors with guidance in how to teach the materials. A comprehensive teachable unit includes explanations about how the materials enGauge a diversity of students and align with the goals.

Unconscious bias
The filters that we all apply unconsciously to situations and people, which can affect student learning and engagement in the classroom.

CHAPTER 2

Active Learning

Active Learning Defined

Active learning implies that students are engaged in their own learning. Active teaching strategies have students do something other than take notes or follow directions, placing the responsibility for learning squarely on their shoulders. As they participate in activities that involve group learning, problem solving, or inquiry-based learning, students construct new knowledge and build new scientific skills. The process of active learning is consistent with scientific teaching principles.

Why Active Learning?

Capturing the spirit of science in the classroom

Sharing the results of an experiment with peers in the laboratory, chatting with a colleague about a recent paper in *Science,* or learning a new experimental technique from a friend are common experiences for working scientists. Rarely is a new idea the product of a single mind. Scientists depend on each other to criticize ideas in seminars and reviews of manuscripts and research proposals. Participation in rigorous, open, scientific debate is one of the most stimulating aspects of being a scientist. Today, most high-impact scientific research is the product of interdisciplinary teams, and the success of these teams is entirely dependent on the ability of the members to work as a group. And yet introductory science courses rarely capture the spirit of dialogue that is characteristic of the scientific enterprise. Students often think of biology as a set of facts that come from a textbook, rather than from a dynamic process that involves experimentation and development and revision of ideas. Active learning is one means of demonstrating that science is not merely a collection of facts, but rather an extended process of probing for answers (Tobias 1990). Conveying the excitement and intellectual challenge of this process is essential to retaining a talented, diverse student population (National Research Council 2003). Participation in a community of biologists is essential to understanding the process of scientific inquiry. Therefore, we must turn our introductory science classrooms into communities of scientists in order to teach effectively what it really means to be a scientist.

Enhancing learning with active participation

Active learning is not a new concept; it is integral to the roots of epistemology articulated by Plato (1901). Socrates, engaging his disciples in group questioning and argument to develop their philosophical ideas, used a form of active learning, and it has endured as a way of learning in some cultures for generations (Dewey 1916, Swisher 1990, Haynes and Gebreyesus 1992, Jagers 1992).

In biology, a number of studies show the benefits of active learning in lectures. For example, Jennifer Knight and William B. Wood at the University of Colorado taught an upper-level developmental biology course in a traditional lecture mode in the fall of 2003 and in an interactive format in the spring of 2004 (Knight and Wood 2005). The interactive format included assigned groups, in-class questions (similar to ConcepTests by Mazur 1996) with electronic audience response systems, group problem-solving activities, journal article discussions, and undergraduate learning assistants. Students started at approximately the same place in both semesters; the differences between the pretest scores were not significant. However, Knight and Wood found that the post-test scores in the interactive course were 9% higher than the previous semester (p=.001). More notably, the interactive course showed a 16% increase in normalized learning gains (using formula from Fagen *et al.* 2002). These results are shown in Figure 2.1.

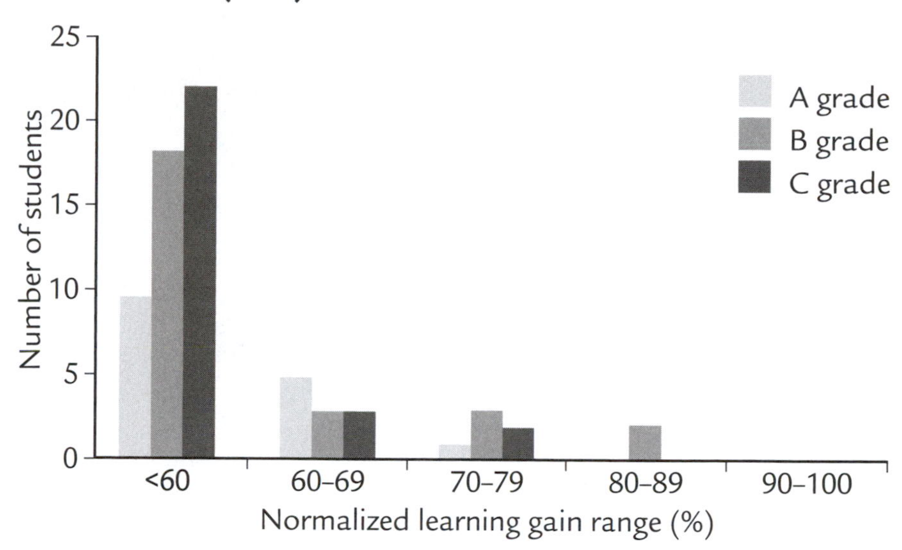

B . Interactive (S'04)

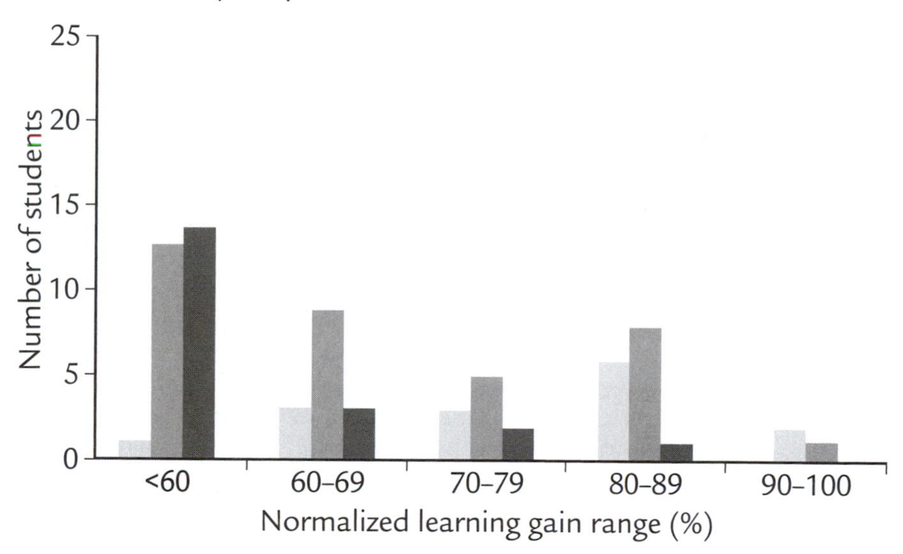

Figure 2.1 Comparison of normalized learning gain ranges (% of possible maximum) achieved by students in each passing grade range ("A," "B," and "C") in the fall 2003 and spring 2004 courses. Normalized learning gains were computed as 100 x (post-test score - pretest score)/(100 - pretest score) (see text). **A.** Fall 2003 (traditional class). **B.** Spring 2004 (interactive class) (Knight and Wood 2005).

In a similar study in an introductory biology course, Dan Udovic at the University of Oregon compared a traditional passive lecture mode with an active format, called "Workshop Biology" (Udovic *et al.* 2002). The differences in learning were quite striking. In a comparison of performance on a test given before and after the course, the students in the Workshop Biology course demonstrated significantly greater proficiency ($p<0.01$) than students in the traditional lecture course (Figure 2.2). Similar results have been achieved in other studies in biology teaching and other fields of science (Hake 1998, Beichner *et al.* 1999, Ebert-May *et al.* 2003). Consistent across these research studies are the findings that active engagement enhances learning and retention, and that active learning builds higher-order thinking skills and reaches a diversity of students.

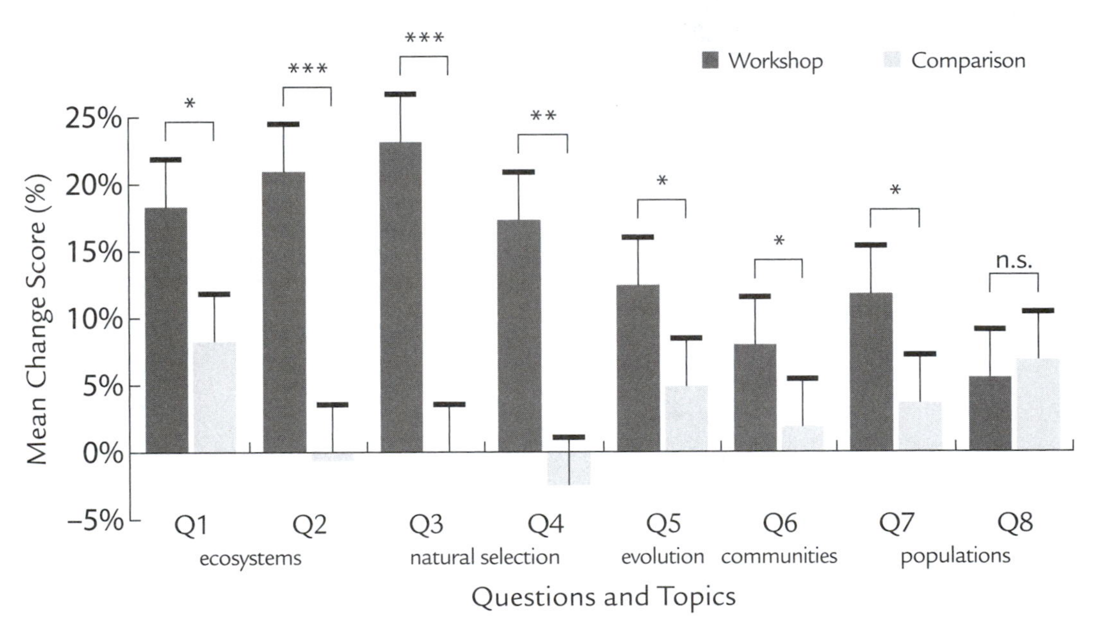

Figure 2.2 Mean change in scores (based on comparing performance on pre- and post-tests) from exam questions in workshop biology vs. traditional lecture. Q1, Q2...Q11 indicate questions included on the pre- and post-tests. Error bars represent one standard error (*$p<0.05$; **$p<0.01$; ***$p<0.001$; n.s. $p>0.05$) (Udovic et al. 2002).

Insight into student learning

Involving students actively in the science classroom enhances understanding and retention and reflects the nature of science (Handelsman *et al.* 2004). In addition, active learning strategies facilitate assessment of student learning. Assessment tools and active learning methods converge, or perhaps are the same thing, because as soon as students become active, they—and the instructor—will discover what they know or what they are confused about. This type of assessment, threaded throughout the teaching process, is a key to good teaching because the instructor and the students monitor learning before an evaluative or summative assessment, such as an exam.

Getting the Most out of Active Learning

Active learning includes a variety of methods. The common feature is that all students in a classroom need to *do* something. It may be quietly thinking, discussing an idea in a group, conducting an experiment, or writing a question or idea. The key is to engage the students so that they are constructing knowledge. Later in this chapter, practical ideas, formats, and tips for active learning will be presented. But first it is worth examining some of the global principles associated with implementing active learning to ensure that students get the most out of the exercises.

Key concepts

It takes time and effort for the instructor to prepare active learning exercises. Utilizing the principle of backward design, the instructor should decide on the learning goals before designing the active learning exercise. The time invested needs to be worthwhile and students need to know that they will be tested on the material that is taught through the active event.

One way to emphasize that the topic of an active learning event is important is to provide students time to reflect on or copy the material they created. For example, if they create a concept map in a group, it is important for them to each have a copy, so time in class to copy the map may be very well spent. In addition to sending the message that this concept is crucial, reflective time also provides an opportunity for reinforcement of the idea, which is a key part of learning. An essential element of active learning exercises is to choose the most important or difficult concepts as the focus for an exercise. In designing active exercises, it helps to remember that every scientific fact was once a question, so anything in science can form the basis for active, inquiry-based learning.

The power of group learning

Active learning can take many forms, including both individual and group work. A vast body of research indicates that learning is different in cooperative and competitive settings, and in groups versus individual modes. In 1949, Columbia University professor Morton Deutsch evaluated the effects of cooperation and competition on the functioning of small groups (Deutsch 1949a and 1949b). Deutsch defined cooperative and competitive groups according to basic differences in their goal structures. In cooperative groups, goals can be achieved by most or all group members; in competitive groups, goals can be achieved by some members but not by all. When he compared the groups, Deutsch observed that the cooperative groups scored higher on assessments of coordination of effort, obligation to participate, attentiveness to group members, diversity of contributions, subdivision of labor, understanding of communication, pressure to achieve, productivity per unit time, quality of product, and orderliness in the cooperative group.

Scientists often argue against group learning, saying that love of competition is a prerequisite to success in science, and that it is therefore appropriate for our teaching methods to capture the nature of competition. But one group of researchers, the Johnson brothers at the University of Minnesota, explored common societal myths about competition—that most human interaction in all societies is competitive; that the use of competition will increase the quality of a student's work; that competition enhances the capacity for adaptive problem-solving; that competition builds character; that students prefer competitive situations; and that competition builds self-confidence and self-esteem. Two decades of research by the Johnsons exploded a number of these myths, demonstrating that the success of even the most competitive activities in our culture, such as sports and warfare, require far more cooperative than competitive interactions. They showed repeatedly that cooperative formats foster learning in diverse settings (Johnson *et al.* 1978, Johnson and Johnson 1985). Evidence has accumulated on the effectiveness of cooperative learning in science. Cooperative learning methods have also been applied successfully in the physical sciences (Smith *et al.* 1991, Beichner *et al.* 1999), mathematics (Dees 1991, Duren and Cherrington 1992), and biology (Okebukola 1986a and 1986b, Lazarowitz *et al.* 1988, Ebert-May *et al.* 2003).

A continuing point of controversy surrounding cooperative learning is whether high-ability students in heterogeneous cooperative groups are penalized by working with low-ability students. Rewey and her colleagues (Rewey *et al.* 1992) showed that cooperative learning can heighten learning among low-ability undergraduate students without diminishing the performance of high-ability students. Physicist Robert Beichner showed in a national study of physics classrooms at three universities that all students benefited from a cooperative, active classroom compared to a traditional lecture, but high-achieving students benefited the most (Beichner *et al.* 1999). Student achievement is highest when 3-5 academically heterogeneous students learn to function together as a team to solve difficult problems over an entire semester. The exception to the "heterogeneous rule" is that, when possible, more than one minority student or woman should be assigned to a group. Research shows that it is advisable to ensure that women and minorities are not sole representatives of their gender or race in a group.

Uncover misconceptions

Misconceptions are inevitable. Instructors and students need to know what students already know and what misconceptions they have about the material, and to use that information to create a classroom where students actively challenge themselves and each other to construct new knowledge (National Research Council 1999b, Modell *et al.* 2005). Left unchallenged, too many graduates leave college clutching the same misconceptions about science they arrived with four years earlier (Schneps 1989). One of the advantages of an active classroom community is that misconceptions are likely to be revealed quickly because most active teaching methods also function as diagnostic tests. The power of this approach is that misconceptions can be dealt with immediately in class, instead of after an exam.

Interestingly, it is far easier to teach a subject to people who know nothing about it than to correct misinformation because simply telling students the answer is rarely effective in fostering conceptual change (Macbeth 2000). If students believe that the foundation of evolution is willful adaptation of individual organisms in response to changing environmental conditions, lecturing them about Darwin's theory of natural selection likely won't change their minds. However, if students have an opportunity to grapple actively with the results of Darwin's observations—in which variation and natural selection are the logical explanations—they are more likely to recognize that their Lamarckian explanations are inconsistent with Darwin's findings. The resulting conflict can serve as the basis for constructing new knowledge.

Enhance human diversity in science

Classrooms that respect, value, and include the contributions of all students will be more likely to attract and retain women and minorities, who often express a sense of alienation, exclusion, and disenfranchisement in the traditional science classroom (Okebukola 1986a, Little Soldier 1989, Hewitt and Seymour 1997). We should challenge ourselves to interest women and minorities in science, to attract them to our courses, and to provide them with a positive environment, the necessary stimulation, and the feeling that they are valued members of our educational community. The traditional competitive, fact-oriented classroom does not promote deep learning for many people; moreover, this learning environment appears to be less effective, on average, for women than for men. Active learning strategies in the science classroom can be readily designed to include women as active contributors within a learning environment of interaction and collaboration. In addition, teachers who have applied these methods have found that what works well for women also works for many men, minorities, and students who typically feel alienated and disempowered in the science classroom (Rosser 1990).

Active Learning in Practice

Many instructors would like to build an active classroom but lack specific strategies to confront the everyday situations and surprises that arise. Many comment that they would like to involve the students in discussion, but the students simply will not talk. Other instructors are afraid of correcting students for fear the students will withdraw. Still others have trouble overcoming their students' feelings of not belonging in science. The following are a few proven tips and tools to engage students and help teachers deal confidently with daily classroom management.

Group problem solving

The simplest method to engage students actively is to present a problem to the class as a whole, instruct the students to consult with the students sitting on either side of them in groups of three students for four to five minutes and then report to the entire class. Each group is then asked to report the results of their group's consultation and the results are recorded on the blackboard. In large classes, the process can be shortened by asking for answers that differ from those already reported. If this approach is used, it is essential to start the reporting process with groups from a different part of the classroom each time the exercise is used so that all groups have the opportunity to report their results over the course of the semester.

Content matters, and group problem solving can help students understand the content better. It's not difficult to take a statement from a lecture and convert it to a group problem-solving exercise. What is harder is deciding which concepts are important enough to warrant the extra attention. Convenient formulations of group problems involve aspects of the scientific method. Students can be presented with an observation about the natural world and then asked to confer in small groups to develop hypotheses that would explain it. Similarly, they can be asked to design an experiment to test a certain hypothesis or interpret experimental data. Another effective version of group problem solving is known as "think-pair-share." Students first think individually for a minute then share and discuss their ideas with a partner. Next, they share their collective answers with the entire class. Table 2.1 provides examples for converting lecture material from passive to active.

Electronic audience response systems

A technological development that has catapulted active learning into common use is the advent of "clickers"—electronic key pads into which students enter answers to a question. Through either infrared or radio waves, their answers are transmitted to a central receiver that is connected to a computer that calculates and displays the frequency distribution of the answers. The responses can be projected for everyone to see. Clickers can provide an immediate measure of the students' conceptual understanding. They offer the opportunity to test understanding before and after an explanation or exercise, enabling the students to assess the trajectory of their own learning. William B. Wood at the University of Colorado (2004) says about clickers:

> Like any technology, these systems are intrinsically neither good nor bad; they can be used skillfully or clumsily, creatively or destructively. However, they can produce results that are eye-opening and potentially of great value to both stu-

Table 2.1 Conversion of lecture material from passive to active.

Concept	Passive Lecture	Active Lecture
Differential gene expression	Every cell in an organism has the same DNA but different genes are expressed at different times and under various conditions. This is called gene expression.	If every cell in a plant has the same DNA, why do different parts of the plant look different? Work with a neighbor to generate a hypothesis.
Differentiation among tissues	Different vascular tissues in plants have different functions. Xylem is mostly a passive transport system that moves water absorbed by the roots. Phloem is a more active system that requires ATP to carry sugars from the leaves and stems to other plant tissues.	In plants, water moves away from the roots to other plant structures through the xylem—often against gravity—but the process does not require ATP like sugar transport in phloem. Generate a hypothesis to explain how this can happen.
Structure of DNA	Complementary base pairing is key to the mechanism of DNA replication.	What do you know about the structure of DNA that suggests a mechanism for replication? Think about it for a minute and then discuss with your neighbor.
Decision making	Many people have concerns about genetically modified organisms. Some of these concerns are well founded and others are not. You have to decide for yourself.	Split the class into two groups for a debate. One group will brainstorm about the potential of genetically modified organisms to be used for beneficial purposes and the other group will discuss possible harm.
Data analysis and interpretation	Based on the data shown in this slide, researchers concluded that *Snarticus inferensis* is the causal agent of the disease.	Consider the data from the experiment I just described. Which of the following conclusions can you draw from these data? Let's take a vote and then discuss the results.

dents and instructors for enhancing the teaching-learning process. . . . The use of clickers can lead to a substantial increase in student active engagement during class . . . to obtain valuable real-time feedback on concepts that students have difficulty with. The give-and-take atmosphere encouraged by the use of clickers in our experience makes the students more responsive in general, so that questions posed to the class as a whole during a lecture are much more likely to elicit responses and discussion.

It is essential that instructors learn the system before using it in the classroom. This technology is still new and students tend to be enthusiastic about it when it works, but they are impatient when technological bugs need to be worked out (Hatch *et al.* 2005). Many other technologies are available for instructional use and are described in DeHaan (2005).

Brainstorming

Presenting a broad, open-ended problem for the whole class to discuss and solve is an effective way to engage students and probe their prior knowledge. This exercise works especially well in a lecture setting and has the advantage of requiring little advance preparation time and no supplies or materials. The ideas generated by the students can serve as the basis for future lectures, which both enhances learning and gives students a sense of contributing substantively to the course content. A modified version can be done with electronic submissions prior to class, so the next class period can begin with a discussion of the brainstorm results.

Examples of Collective Brainstorming Challenges

- Imagine you are a cell. What are your greatest challenges?
- Suggest new uses for genetic engineering in medicine or agriculture.
- Think of some examples of natural selection.
- What are the similarities or differences between . . . and . . .?

One-minute questions

Asking students to write short answers to questions at the end of class offers students an active role in learning and a quick way to assess learning. Because the answers to the questions are short (short enough to fit on a 3 × 5 inch index card), they provide a way for students to gain regular writing practice without adding a huge reading burden for the instructor.

A class routine can be established so that students pick up class handouts and an index card on their way into class. They drop off their index cards in boxes marked with their lab or discussion section on the way out of class. Whether or not the cards are graded, the students should be asked to put their names on them to encourage accountability.

One-Minute Questions

Reflection on learning

- What concept presented in class this week was difficult for you?
- What was the key concept in today's lecture?
- What else would you like to know about today's topic?

Critical thinking

- Describe the connection between the content of today's lecture and your life outside the classroom.
- Describe how your own personal bias, shaped by your background, ethnic origins, culture, experience, religion, education, or gender might affect your interpretation of the material presented today.

The answers to these questions can be illuminating to both students and teachers. The simple act of reflecting on them and writing down the answers may lead students to realize that they should review the concept or see the instructor for help. A large number of students citing the same point of confusion may likewise lead the instructor to recognize the need to return to the concept during the next class meeting. Other questions can encourage critical thinking and show students the relevance of biology and its connections to their everyday lives.

Strip sequence

A class activity to strengthen students' logical thinking processes and test understanding of biological or physical processes is the strip sequence. Step-by-step processes lend themselves well to this technique, but other material can also be used. Students are given the steps in a process on strips of paper (or the virtual equivalent) that are jumbled (Figure 2.3). The challenge for the students in class is to work together to reconstruct the proper sequence. Making logical connections helps the students dissect and reconstruct the process, improving understanding. Alternative versions of strip sequences use pictures or involve students in creating the strip sequences from a textbook.

Cut into strips on lines, mix the strips, and instruct students to reassemble them in the correct order.
The first step in the process of gene expression is transcription.
In this step, DNA is used as the template for synthesis of RNA (mRNA).
Base pairing between one strand of DNA and RNA bases, following the rules of base complementarity, defines the base sequence of the mRNA.
The enzyme RNA polymerase is required for mRNA synthesis.
In the next step, known as translation, mRNA bases pair with transfer RNA molecules (tRNA).
Each mRNA contains many 3-base units called codons; each tRNA has a unique 3-base unit called an anticodon.
Each tRNA carries a particular amino acid.
As the tRNAs line up along the mRNA in the order defined by codon/anticodon recognition, they define the sequence of amino acids in the protein.
Peptidyl transferase detaches the amino acids from their tRNAs and links them together to form a protein.
A string of amino acids makes up a protein, and proteins give an organism its distinguishing characteristics.

Figure 2.3 Strip sequence on gene expression

Decision making

Acquiring the power to make decisions that affect the lives of others and being held accountable for those decisions is a strong incentive to ask questions and to learn and evaluate facts about an issue. Students are asked to imagine that they are policy-makers who must make tough decisions that require scientific information. The desire to appear responsible and rational induces them to become experts on the issue, which will require learning information, thinking critically, and developing a creative solution.

Examples of Decision-making Activities

- You are the director of the antibiotic discovery unit in a major pharmaceutical company and you are asked for a five-year plan to develop new antibiotics. You are told that the plan will be funded only if you can convince your managers that you will be able to develop five new drugs with entirely new modes of action. Can you do it? What is your plan and how will you defend it?
- You are the head of a major blood bank, and there is a worldwide blood shortage. You are offered a shipment of blood that might be contaminated with a new retrovirus that has not been well studied. Will you allow the blood to be used? Why? What would you like to know before you make your decision?
- You are the Executive Director of the National Pesticide-Free Food Network. You hear that a chemical company is about to register a new fungicide. The fungicide reduces accumulation of aflatoxin, a highly carcinogenic toxin produced by a fungus that often grows on peanuts. You must decide whether your group should protest the use of the fungicide on peanuts. What questions will you ask and what process will you use to decide whether or not to fight the use of the fungicide?
- You are the ecologist for a land conservation trust, and your most recent project is to restore a degraded oak savanna ecosystem. As a part of your management activities, you will need to remove many of the trees and shrubs that have invaded the savanna. However, you know that many of the local landowners will have strong objections to cutting down any of the trees. How will you address the neighbors' concerns? What ecological concepts will you use to justify your plan?

Concept mapping

Concept maps are graphic representations of the relationships between concepts. The validity of each proposition is based on the two concepts that are associated, the description that the student provides to describe their relationship, and the direction of the association.

Steps for Concept Maps

1. Identify the key concepts to be mapped.
2. Determine the general relationship between the concepts and arrange them two at a time on a piece of paper. The proposed relationship between the concepts is the "proposition."
3. Draw an arrow from one concept to another and then describe the relationship in a short phrase next to the arrow. The relationship should be read in a sentence in the direction of the arrow. For example, two concepts, "chloroplast" and "chemical energy," are joined by the phrase "produces." The arrow would be drawn from chloroplast to chemical energy, and the phrase would read, "Chloroplast produces chemical energy."

Since they were first developed in 1972 at Cornell University, concept maps have been shown to enhance learning and development of higher-order thinking skills (Novak and Canas 2006). Concept maps or other techniques that require students to reformat material are essential elements of a classroom that reaches diverse learners because in reformatting, individual students construct a framework that makes sense to them. To do so, they must grapple with the material and understand it. Students should have the opportunity to try out different formats and choose the format that suits both the material and their own learning styles. In addition, concept maps illustrate that there can be many different ways to represent the relationships among concepts, underscoring the value of a diversity of approaches to thinking in science.

Concept maps redux: Mini-maps

Concept maps make evident to students the diversity of approaches to complex problems. Mini-maps embrace the idea of concepts maps, but they are much simpler to execute in the classroom. Their advantage is that they can be done in nearly any type of classroom—including large lectures—with concepts at any level of difficulty. In addition, they provide immediate feedback about what students understand. Students are free to discuss their reasoning and work with instructors and classmates to identify misconceptions as they come to consensus about how the terms relate to each other.

Mini-maps are graphic representations of a concept or idea, using a set of related terms (usually 10 or fewer).

Steps for Mini-Maps

1. The instructor provides a list of terms, which can include major concepts, specific terms, or even a red herring that doesn't belong. Students write the terms on individual pieces of paper or notecards.
2. Students work in groups of 3–5 to arrange the terms in a logical structure. The relationship between terms needs to be explained with directional arrows that include a word or phrase.
3. While students are developing the mini-map, instructors interact with the groups and ask students to articulate their logic in determining the relationship among terms, or ask a few groups to explain their map to the rest of the class.

Cases and problem-based learning

Case-based and problem-based learning present students with "real-life" situations or examples to which they must apply their factual or content knowledge. Both are designed to promote higher-order critical thinking more than factual recall. These methods demonstrate the relevance of course content to situations beyond class contexts, making facts and concepts easier to learn and retain.

The differences between problem- and case-based learning, however subtle, are meaningful, and understanding them can help instructors choose the appropriate method. Problem-based learning poses a specific question or puzzle that must be solved in order to reach a particular goal (Allen and Duch 1998, Duch *et al.* 2001). As part of solving these puzzles, students must first assess what they already know and determine what in their arsenal of knowledge is relevant or useful in solving the problem. The complexity of the problem itself becomes the means of organizing learned material, thus reinforcing students' knowledge and bolstering their confidence in their understanding of course material. The limits of the students' knowledge becomes the basis for further learning because students identify areas in which they need additional information to solve the initial problem, locate sources that may be of use, and integrate the new information with what they already know. Problem-based exercises therefore encourage self-evaluation and research along with factual recall and application.

Case-based learning also presents specific, real-world situations that students must evaluate before applying their knowledge and formulating an answer (Waterman and Stanley 2005). In contrast to problem-based scenarios, however, case-based situations and questions are open-ended. They typically do not have one single correct answer, and the complexity of the answers (not the problems themselves) becomes the means by which students assess and evaluate their understanding of both the course content and the case at hand. In case-based learning, students are asked to make a decision based upon the description of the case and their existing knowledge. Students must decide if they have enough information to make an informed decision, but also must consider what broader impact their decisions may have. What issues are at stake in the case at hand? Who might be affected—either positively or negatively—by the decision, and how might that knowledge influence the final outcome? Such questions require students to think beyond the immediate context of the case (*e.g.,* the course setting and their grade) and demonstrate how scientific knowledge is an essential part of creating policy or resolving complex issues.

An exception to the open-ended cases are diagnostic cases, used frequently in health and medical fields. For example, Harvard Medical School has used online case studies in its toxicology courses. Students are asked to make a diagnosis or choose a treatment plan based on a battery of vital statistics, patient histories, and diagnostic cues. While these cases often have one correct answer, they present students with critical, constructive explanations of answers both incorrect and correct.

Management for the Active Classroom

Start early

It is important to set the tone early in the semester. If the students receive the clear message that they are expected to be active members of the classroom, and if they perceive immediately that their ideas shape the classroom, then a community spirit will naturally follow.

Start simple

It is useful to start with simple, open-ended, or familiar topics the first time active learning exercises are used. Students may be afraid of being wrong or feel convinced that they know nothing about the subject if this is their first science course. If the first question they are asked is hard or has only one right answer that they may not know, they may not be eager to participate. The challenging questions should be saved for later when the students have more confidence or have learned that it is acceptable to be wrong. The first question could be something that makes a connection between the course and the rest of their lives.

Questions to Connect the Course to Students' Lives

- How does science affect your life?
- Describe something you would like to understand about the natural world.
- What issues facing the world today involve science and technology?
- What do you hope to learn in this course?

Groups can be asked to set objectives for the course and report back to the entire class. One student can be designated to record the list and the class can return to it every few weeks with the class to determine whether the students' objectives are being met and whether new objectives have emerged.

Make it personal

Students strive harder and achieve higher standards if they are thought of as individuals and not just identification numbers. It can be difficult in large introductory courses to make personal contact with students, but a few tricks can help. Pictures of the students in groups with name cards on can help instructors and students learn names. Instructors can call on students in class by name even if the instructor knows only a few names; the classroom will have a more personal feel and the students will feel more committed to the class. Alternatively, when the students speak in class, they can be asked to say their names or hold up file folders with their names written in large letters. Then their ideas can be referred to with their names attached. Students hearing an instructor say, "as Heather pointed out," will develop pride in their ideas and a sense of being valued as members of a scientific community, which are important ingredients in maintaining active engagement.

Even if it is impossible to remember all of the students' names, instructors can let them know that they are noticed by addressing individuals with questions or comments. Chatting with a few of them individually for a minute before each class or asking certain students why they missed the previous class, why they sat in a different seat last time, or whether something said in lecture was clear and understandable can develop a critical rapport. These actions are easy, require little time, preparation, or follow-up but have a powerful impact on the willingness of students to participate in a class. Students are often astonished that a teacher would care about them and their opinions. Showing that scientists care about people and their ideas will make science seem more human and interesting.

Take risks

Sometimes students simply will not respond to a request for ideas. They may be too inhibited or they may not quite understand the material or the question asked. Often, gridlock can be broken by proposing an idea for the students to analyze or evaluate. Proposing an obviously silly experiment and asking the students if it is going in the right direction can jump-start discussion. For example, if students have been asked to design an experiment to learn about the proteins in muscle cells, the instructor might suggest starting with grinding up a liver and ask them if that is the right place to start and why. This approach can be humorous, but it serves the much deeper purpose of having the students realize that they know something about the topic at hand even if it is simply that the liver is not a great source of muscle cells. It is also valuable for students to learn to criticize ideas from any source, including the instructor.

Give positive feedback

One of the simplest approaches to promoting participation is to be encouraging. This does not mean accepting students' ideas without scrutiny. The key is to validate them as people while honestly evaluating their ideas and challenging them to evaluate their own and others' ideas. Students respond powerfully to trivial comments. Simply saying, "Good question!" before answering a question can put students at ease and encourage them to ask questions freely. It is important to avoid words and behavior that make students clam up. Using harsh phrases such as "You would know the answer to that if you had done the reading" or making jokes about a particular student's grade in front of the class is bound to make students hesitant about contributing to the class. Instructors often don't realize how offhand comments or jokes can affect the engagement of students in the class.

Deal with "wrong"

Wrong answers are a necessary part of uncovering misconceptions. Although dispelling misconceptions takes much more than simply telling students the right answer, it is essential that incorrect information be corrected so that these ideas are not incorporated into students' thinking. However, students who have expressed wrong ideas have exposed themselves by sharing their ideas with the class, and it is important not to embarrass them or make them feel personally attacked. In managing an active classroom, it is essential to keep criticism of ideas separate from criticisms of people, and to foster an environment where there are no "dumb" questions. In fact, the student who brings up an idea that is simply wrong according to current scientific knowledge may do a great service for the class and can be supported for doing so. By thanking the student for being brave enough to volunteer the idea, the instructor can provide personal support and respect for the person while making absolutely sure the idea is thoroughly evaluated and corrected. It may be possible to show the merits of the idea while correcting it: it may have historical value (if one of our students believes it, there was probably once a debate in which a prominent biologist defended the idea), it may be instructive as a contrast to the right idea, or it may simply be a common misconception that needs to be addressed. A combination of approaches can be used to defuse the embarrassment of being wrong and to ensure that the correct information or idea is firmly established.

For example, let us imagine that a student named Horton states that proteins are the universal hereditary material. Hearing this, most instructors would panic and wonder what they did wrong to instill such a wrong-headed idea, but once the panic subsides, the instructor could use a number of approaches to arrive at the concept that nucleic acids are the hereditary material in biology. First, the other students might be asked if they agree with Horton's statement. This approach encourages the students to be critical of each others' ideas and provides the opportunity for Horton to be corrected by a peer instead of by the instructor, which may be less intimidating to him. If the other students do not correct the idea, then make the point that this is clearly a widely held misconception and it is a good thing Horton brought it up. This shows support for Horton and calls attention to the fact that many of the other students probably need to learn the information that is about to be presented. Whether a student or the instructor corrects the misconception, Horton's pride can be protected by discussing the fact that although all current evidence shows that proteins are *not* the universal hereditary material, there was a substantive debate about the chemical nature of hereditary material involving many prominent biologists in the early part of the twentieth century. Finally, Horton or anyone else in the class can be invited to imagine why nucleic acids might make better hereditary material than proteins, or how cells would function differently in transmission or expression of genes if genes were proteins. This takes the emphasis off the "wrongness" of Horton's idea and turns it instead into the basis for a creative, educational exercise. Thanking Horton in or after class for contributing the spark for this discussion makes it clear that a part of science—indeed, an essential part—is experimenting with ideas and using discussion to arrive at a correct answer.

Maintain control and high standards

It is important to maintain control and insist on high standards of thought and behavior. It is essential to prevent a few students from dominating or letting the class become a free-for-all. The students need to know that they cannot present sloppy or wrong ideas without being challenged. For active learning to be successful, two things must be absolutely clear: in the classroom, every idea will be rigorously evaluated, and it is okay to be wrong.

Ethics of group work

Encouraging or requiring group work will act as an invitation to some students to cheat, perhaps by copying another student's assignment, using lab results from an experiment they did not perform, or simply by not contributing to the group effort. The best method to prevent this is to make expectations clear. Cooperative work provides the substance for useful discussions about ethics in science, giving and sharing credit, and the difference between fair exchange and theft or exploitation. To prevent abuses of the system, the rules should be made explicit by being written down in course handouts. A separate handout on ethical conduct, explaining what is acceptable and what is not, is very much appreciated by students. The instructor should consider asking students to sign a contract for ethical conduct that states the requirements of the classroom (see Figure 2.4). Committing in writing to uphold ethical standards is probably enough to prevent most students from cheating. If instructors find that students cheat, this is not a reason to give up on these methods; the same teaching methods that made the class vulnerable to misconduct by a few enhance learning for many students.

Design the classroom to be inclusive

Expecting students to work together requires that they are near each other (Beicher *et al,* 1999). In many lecture halls, this may require a little reorganizing, but it's usually possible. The best classrooms integrate mobile chairs and tables that accommodate flexible arrangements and facilitate discussion. The most important component of an inclusive classroom, however, is the expectation that students will work together respectfully. If learning experiences and instruction are designed to motivate students to interact, physical classroom design can often be overcome.

Design the out-of-class experience to be inclusive

Much of student learning occurs outside the classroom. It's important to influence positive group dynamics and inclusiveness there as well. Most instructors would probably object to racist behaviors or bigoted comments in their classrooms. But prejudice can manifest in more subtle behaviors, such as exclusion of certain students from study groups, which can be damaging, especially when experienced repeatedly throughout college courses. The spirit of inclusivity that is essential to learning inside the classroom walls should pervade the out-of-class experience as well. Homework assignments and out-of-class projects should be structured to encourage students to draw on each other's knowledge and understanding, and the expectations for collaboration should be made clear.

Expectations for Students Participating in a Cooperative Classroom

Learning in a cooperative environment should be stimulating, demanding, and fair. Because this approach to learning is different from the competitive classroom structure that many other courses are based on, it is important for us to be clear about mutual expectations. Below are my expectations for students in this class. This set of expectations is intended to maximize debate and exchange of ideas in an atmosphere of mutual respect while preserving individual ownership of ideas and written words. If you feel you do not understand or cannot agree to these expectations, you should discuss this with your instructor and classmates.

1. Students are expected to work cooperatively with other members of the class and show respect for the ideas and contributions of other people.
2. When working as part of a group, students should strive to be good contributors to the group, listen to others, not dominate, and recognize the contributions of others. Students should try to ensure that everyone in the group makes a contribution, and recognize that everyone contributes in different ways to a group process.
3. Students should conduct experiments, discuss group exams, and develop projects as part of a group, but write lab reports, exams, and papers alone and not copy from anyone else. If you use material from published sources, you must provide appropriate attribution.

I have read and understood the expectations of students in this class. If I am uncertain about appropriate behavior in the class, I will ask one of the instructors for clarification.

Signed,

____

Please print your name here

Keep one copy for yourself and return the other copy to your instructor.

Figure 2.4 Sample contract that states classroom expectations for ethical conduct.

CHAPTER 3

Assessment

Assessment Defined

Assessment is a tool for understanding what students are learning. Many instructors know intuitively that assessment is integral to teaching—that students learn from preparing for and taking exams. The philosophy of this chapter expands upon that knowledge. If students learn from taking exams, and we discover what they have learned by grading the exams, then why wait for the exam to test them? Both students and instructors benefit from the results of regular, ongoing assessment when it is used to "promote and diagnose" learning (Huba and Freed 2000). In short, "assessment is more than grades . . . it is feedback for students and instructors . . . and it drives student learning" (National Institute for Science Education 1999a).

Why Assessment?

Assessment should seem natural to scientists because it parallels parts of scientific research. In science, data collection is central to discovery; in scientific teaching, instructors collect data to evaluate teaching and learning. The information generated by assessment is crucial because it informs change, making assessment the fulcrum of scientific teaching. Assessment is omnipresent in articles about undergraduate science education; journals like *CBE-Life Sciences* and *BioScience* explicitly include assessment among their submission requirements. Furthermore, recent publications demonstrate that scientists have successfully integrated assessment principles into their teaching (Udovic 2002, Ebert-May 2003, Knight and Wood 2005).

The literature about assessment includes a breadth of accessible and useful resources. In 1993, Angelo and Cross compiled *Classroom Assessment Techniques: A Handbook for College Teachers*, a compendium of teaching methods and examples for incorporating assessment into college classrooms. Six years later, Huba and Freed (2000) wrote *Learner-Centered Assessment on College Campuses: Shifting the Focus from Teaching to Learning*, an assessment resource that explicitly focuses on learning. These two resources, among others, have guided how college instructors approach assessment for the past decade.

Ongoing assessment increases learning gains

Assessment has the obvious purpose of *monitoring* learning, but the consequences are more profound from the student perspective when it is used to *promote* learning. In an extensive review of research about assessment in the classroom, Black and Wiliam (1998) concluded that ongoing assessment plays a key role—*possibly the most important role*—in shaping classroom standards and increasing learning gains. They reported that well-designed, regular assessment of students had more impact on student learning than any other educational intervention. In addition, they found that high-caliber formative assessment increased learning gains for all students, but it had the most impact for low-achieving students. According to Black and Wiliam, "formative assessment . . . is at the heart of effective teaching." Thus, assessment—the process of determining progress toward and achievement of goals—is an essential component of quality instruction.

> "Ongoing assessment plays a key role—possibly the most important role—in shaping classroom standards and increasing learning gains."
>
> Black and Wiliam, 1998

Assessment tools that provide regular checkpoints and measures of achievement let the students determine whether they are on track and accordingly modify their approaches. Specifically, regular, ongoing assessment provides a mechanism for students to evaluate themselves and each other. As a result, learning becomes a process of reflection and analysis with specific markers of achievement, rather than simply an end point and a grade. The resulting information helps guide changes in student study and learning behavior.

Assessment provides feedback to students and instructors about learning

According to Wiggins and McTighe (1998), there are two important features in assessment: (1) what kind of performance or behavior indicates understanding, and (2) what specific criteria differentiate the levels of understanding. In other words, assessment relates to two aspects of students performance: *what* students do and the *caliber* of their performance. The part that students do is simply the activity in which they are expected to participate. The caliber of performance deals with how well the students perform the activity. For a multiple-choice question with a single, correct answer, this is clear: A correct answer indicates excellent performance. However, more complex projects that involve writing or presentation can be trickier to evaluate. In this case, it helps to have a mechanism for defining excellence. Rubrics can be powerful tools to help students achieve excellence, and to keep instructors focused on their goals.

The primary feature of assessment is that it provides feedback to instructors and students about learning and teaching. When assessment is integrated into the learning process, students learn to differentiate between what they already know and what they need to learn, which helps focus and motivate learning. Assessment is typically categorized in two ways: formative and summative. Both formative (during the teaching event) and summative (at the end of the teaching event) assessments offer information about student learning that can shape learning behaviors and guide instructional decisions.

Feedback to students

When assessment is routinely integrated into the curriculum, it provides a mechanism to engage students and shape their learning behaviors. In Chapter 1, we described the importance of **metacognition**—the ability to carry on an internal dialogue about what is being learned. Assessment allows students to gauge their own progress toward the learning goals and provides the feedback they need to prompt changes in their study habits. Feedback from the assessment activities, therefore, becomes an integral part of the learning process instead of just a checkpoint at the end of a unit or semester.

Feedback to instructors

From the instructor's perspective, assessment data should guide changes in instruction, curriculum, and teaching behaviors. Effective assessment informs instructors how students are progressing toward learning goals **while the learning is occurring.** The feedback from assessment guides mid-course instructional changes that can help redirect students toward the learning goals. In addition, assessment tools provide more than grades; the results can promote dialogue between students and instructors and guide changes in instructional materials and teaching. As one educator explained, "For teachers to be effective in achieving learning goals, they must engage in an ongoing process of aligning the content, themselves, and students in a specific context" (Wulff 2005). Whether it's a prequiz about prior knowledge, a homework assignment, a midterm exam, or an in-class activity, the time the assessment takes should be consistent with the relative importance of that knowledge or skill set as a learning goal.

Simple assessments can help guide instructional decisions. For example, a brainstorming activity can elicit students' prior knowledge about a topic through a series of questions that students answer with an audience response system ("clickers"). If the results indicate that most students already know the topic, then the instructor may elect to skip the topic, probe deeper to determine how much they know, conduct a brief review, or delve into an application of that topic that requires more complex analytical skills. If the material is new to the students, or they have misconceptions about it at the beginning of class, then the instructor might consider revisiting the same question(s) again at the middle or end of class to see if their understanding has improved. If understanding does not improve, the instructor should consider why and decide what teaching action to take.

Activities can likewise provide qualitative data if students have the chance to convey what they feel is most effective, what they like/dislike, whether they feel the learning objectives are being met, if they like working in groups, whether they struggle to understand, or how the instructor might improve teaching and learning. Feedback from these types of questions can help quantify some of the more intangible aspects of teaching and serve as a mechanism to improve instruction. Moreover, students' investment increases when they are given a chance to make decisions about their own learning.

Assessment and diversity

It is important for the instructor to cultivate an environment in which all students have ample opportunity to gauge their progress toward the learning goals **during the learning process** (National Institute for Science Education 1999a). In addition to increasing learning gains, assessment can help foster inclusive classrooms. Shifting the emphasis from criticism to constructive feedback can foster open dialogue in the classroom. This creates a classroom climate that is respectful and welcoming, yet clearly focused on learning. In addition, feedback from a variety of assessment methods can help a diversity of students take responsibility for their own learning in their own ways. The effect is to catapult learning beyond facts and figures and to create an inclusive classroom where students come to understand the complexities of science as well as the process of learning science.

A word about grading

The reality of most courses is that, at the end of the semester, instructors need to assign a grade. This is where *assessment* becomes *evaluation*. While assessment activities provide the instructor and student with ongoing feedback about student understanding, a second type of assessment—summative assessment—measures progress at defined points in the semester. A well-designed assessment plan should lend itself readily to grading. Students should have a clear vision of what the instructor expects in terms of knowledge, skills, and performance, and they should have a fair amount of experience assessing their learning (and that of their peers) at many points during the course. Summative assessment helps gauge how well the students have learned the material and supplies data that can be converted into a grade. Examples of summative assessments include any of the classic midterm or end-of-semester activities: comprehensive exams, final oral presentations, or poster symposia. The important point to remember is that assessment is *more* than grading; it offers ongoing feedback to students and teachers alike.

Assessment in Practice

"EnGauge" students in learning

Active learning exercises—such as those described in Chapter 2—and assessment tools converge. It is hard to imagine an active learning exercise that does not have an assessment component, and assessment is automatically active because the students must **do** something to assess themselves or be assessed. Many assessment tools, therefore, resemble active learning activities and accomplish similar goals. It is worth considering them in both contexts to understand their multiple roles in teaching and learning.

We have coined the term "enGauge" to capture the spirit of scientific teaching: students simultaneously ***engage*** in learning and ***gauge*** what they are learning. Engaged students are more motivated to achieve the learning goals and take responsibility for learning (Marzano 1998), which is precisely the type of academic curiosity that effective science courses aim to awaken (National Research Council 2003).

A well-designed enGaugement motivates all students to learn and provides instructors and students with feedback about learning. Because it integrates the three core themes of scientific teaching—active learning, assessment, and diversity—an enGaugement can simultaneously engage students and gauge their learning. EnGaugements are particularly effective at addressing difficult concepts or skills, targeting common misconceptions, or emphasizing important points. Many enGaugements also lend themselves well to grading. Table 3.1 provides a list of "enGaugements."

Enlist students in teaching each other: Reading assessments

Reading assessments are one of the most comprehensive examples of scientific teaching, because they target diversity, active learning, and assessment at many levels. The previous section provided ideas for instructors to "enGauge" students. Reading assessments take enGaugements one step further by enlisting groups of students to design the activities and teach each other. At different points in the course, each student group designs and leads an activity about an assigned reading. The goal of each reading assessment is twofold: to engage everyone in learning and to determine whether everyone in the class understands the reading.

Reading assessments are inclusive by design. They give students freedom to design an activity that makes sense to them—and their individual learning styles and preferences—within the constraints of format and time. Students work in groups toward a common goal, so they get all the benefits of cooperative learning, peer teaching, and diverse group dynamics. Over the course of a semester,

Tools to EnGauge Students

Brainstorming
List as many answers as possible to a question.

Case studies
Solve a problem or situation in a real-world context.

"Clicker" questions
Answer questions electronically in class.

Decision making
Work together to recommend solutions to a problem.

Group exams
Work together to discuss exam questions but write answers individually.

One-minute papers
Write a short answer about a topic or question.

Pre/post questions
Answer questions before and after a topic is taught.

Strip sequence
Arrange a series of events into the correct order.

Think-pair-share
Think about possible answers to a question individually. Discuss them with a partner and come to consensus.

Table 3.1 Examples and Objectives of EnGaugements

Biology Example and Instructions	Objectives
Brainstorming Answer the following question in large group. One person records answers. Optional: Arrange the list into two or more categories (*e.g.*, abiotic vs. biotic factors) *Question*: What does a plant need to survive?	Brainstorming elicits responses from large audience and aggregates them into a single list. It provides the instructor and students with an overview of the group's collective knowledge. By separating the brainstorm list into two or more categories, students evaluate how well they understand the role of each response in a specific context.
Case study and decision making Read the following case. Write a paragraph to explain what the patient should do next. Justify your recommendation with biological reasons. *Case*: A patient expressed eye irritation, which the doctor diagnosed as conjunctivitis. Antibiotic treatment alleviated the symptoms within a few days, but the symptoms returned two weeks later. The doctor recommended taking antibiotics again.	Cases engage students in solving a problem in a real-life context. To solve them, students need to evaluate what they know about infectious disease, causal agents, and antibiotic resistance; apply that knowledge to the case; and determine what additional information is needed to make a recommendation.
"Clicker" questions Answer the following question on your electronic response keypad. *Question:* Which organisms are most distantly related? (a) bacteria and archaea; (b) plants and animals; (c) plants and fungi; (d) humans and fungi.	Clicker questions require students to gauge whether they understand a concept or topic, thereby engaging students in the ensuing activities (*e.g.*, lecture) about that topic.
Group exams Work with a group to discuss the following statement. Write your answer individually. *Statement:* Explain the role of aflatoxin in liver cancer.	Group exams engage students in working collaboratively to identify creative solutions to a problem. Writing individual answers requires students to evaluate how well they understand the topic and its underlying concepts.

Table 3.1 Examples and Objectives of EnGaugements (continued)

Biology Example and Instructions	Objectives
Mini-map Arrange the following terms in a logical order. Explain (using arrows or words) how the terms relate to each other. *Terms:* tRNA, DNA, protein, mRNA, amino acid, translation, transcription, replication, promoter	Mini-maps engage students in developing a non-verbal representation of a concept. The process of developing a visual arrangement requires students to evaluate different ways that terms can relate to each other and to appreciate that a biological process may not be unidirectional or linear.
One-minute paper Write for one minute to answer the following question. *Question:* What about the structure of DNA suggests a mechanism for replication?	One-minute papers engage students in articulating their knowledge about a topic or applying their knowledge to another situation. By writing their answer in one minute, students need to evaluate the most important and relevant components of their argument.
Pre/post questions Write for one minute at the beginning and end of class in response to the following statement. Explain any differences between your responses. *Statement:* Describe two mechanisms that a bacterium can use to harm a plant.	Pre/post questions can take many forms, including one-minute papers or clicker questions. They engage students in thinking critically about a specific question or problem. By comparing pre/post responses, students evaluate whether and why their answers changed during the class period.
Strip sequence Use your textbook as a guide and work with a partner. You write the important steps in meiosis; your partner writes the important steps in mitosis. Cut the steps apart and scramble the order. Each of you should try to put the other person's steps into the correct order. Discuss.	Strip sequences engage students in recognizing cause and effect and in determining the logical sequence of events. When students derive their own strip sequences, they need to evaluate the critical steps in the process.
Statement correction Discuss with a partner what is wrong with the following statement. Propose an alternative statement that is correct. *Statement:* "I don't want to eat any viruses or bacteria, so I refuse to buy foods that have been genetically modified."	Statement corrections engage students in evaluating what concepts are misrepresented and in determining what information they need to correct it.

the diversity of activities will illustrate many ways to approach a complex topic or difficult reading.

The nature of reading assessments is active and collaborative. Reading assessments require cooperation by the students who develop and teach them, and everyone benefits from the active-learning environment when they are engaged in the activity during class. Reading assessments also foster deep learning. The presenting groups have to employ several high-order thinking skills to develop an activity. They evaluate which aspects of the reading are most important, choose which topics to include, and design an activity to assess and engage their classmates' understanding of the reading.

Finally—and arguably most importantly—reading assessments gauge how well students understand the reading and its concepts. The instructor can use the feedback to grade the presenting group or as formative assessment about how well students understand the reading topic. The feedback can provide information for students as well. Everyone in the class has a chance to question their own understanding of the reading through the activity, and the presenting group will have had to determine whether they understood the material in sufficient depth to teach it.

One of the best outcomes of reading assessments is that students actually do the assigned readings (and every instructor knows how difficult it is to enforce reading assignments!). If students choose not to read before class, they will not be able to participate in the discussions, and they will let their peers down. It is therefore important that the readings are difficult enough that students struggle with them and come to class with questions about the readings and are ready to engage in learning more.

Many instructors have reservations about turning over so much control to the students. What if the student presenters say something wrong, and the whole class graduates from college thinking that DNA is always single-stranded? Part of the power of reading assessments is they are designed to uncover misconceptions, and if one student is thinking it, many others probably are too. The odds are very good that someone in the group will catch the mistake and correct it because students are working in groups. This can also be a teachable moment: an opportunity to correct the mistake for the entire class. (Chapter 2 has more suggestions for how to deal with "wrong" with empathy and panache.)

Finally, no matter what activity the students design, every instructor should have a back-up plan and a little bit of summary time to discuss alternative views on the main points of the readings, address important information that was not included, underscore key concepts, or correct any remaining misconceptions. However, this time should **not** be used to cover every bit of content from the reading. Remember that the students should have already read it and have just spent time discussing it. This time should be used to ensure that students understand the key concepts; they can use the reading to fill in the details.

Measure both progress and outcomes

Assessment measures both progress and outcomes, which ties closely to the learning goals—what students will know, understand, or be able to do. If a topic, skill, or behavior is important enough to represent a learning goal, then students' progress toward the goal should be assessed. Assessment reinforces each goal's importance and provides students with a framework for gauging progress toward achieving it. Alignment is the key to reducing curriculum clutter and alleviating the nagging sense that you have to "cover everything." To do this, assessment must align with learning goals and communicate clear standards of "excellent" performance. In sum, assessment engages students in learning and enables both students and instructor to identify gaps, confusion, mistakes, and progress. Close alignment of assessment and learning goals helps simplify decisions about what to include and what to jettison, and to set clear expectations.

In Table 3.2, the activity assesses knowledge at three key points: at the beginning of the lab (students have to recognize what they **don't** understand about bacterial cells and growth when they make a prediction), during the lab (students have to decide what to measure and record), and at the end of the lab (students have to explain what a colony is in order to explain whether their evidence supports the hypothesis). In addition, this assessment can be used to test several other learning goals, such as the ability to design a controlled experiment with replication, to understand the concept of microbial diversity, or to work collaboratively in groups.

Give exams that foster learning

One of the most surprisingly effective ways to assess students is group exams. It is surprising because most of us have been taught that exams must be individual,

Instructions for reading assessments

Work in groups of 2-4 students to develop a 20-minute reading assessment based on the assigned reading. A reading assessment replaces lecture, so the activity needs to engage everyone in the class, give context to the reading in relation to the course, and give everyone time to learn actively. On the assigned day, your group will lead the activity. Everyone in the group should be involved in development of the reading assessment and in teaching it in class. Some examples are listed below.

Each reading assessment should:

- Engage everyone in the class in an activity based on the assigned reading
- Determine how well each student understands the main concepts and topics from an assigned reading
- Apply the reading to the relevant topics in this section of the course
- Be creative

Sample Reading Assessments

1. Write three questions that you have about the reading. Discuss solutions in small groups for five minutes. Each group will report one question and its solutions to the large group.
2. List two things you learned from this reading. Explain how each could be applied to another topic in this course.
3. State what you think were the author's main points.
4. Explain how you could use a scientific method or technique from the reading.
5. Answer a set of questions that highlights the most challenging aspects of the reading (in a worksheet or quiz format).
6. Identify which terms are integral to understanding the main points of the reading. Sketch and describe the relationship among the terms.
7. Compare and contrast the ideas in the reading with a manuscript, book, data set, or website.
8. Answer the following questions. What hypothesis or hypotheses were the authors testing in this paper? What evidence did they provide to support or refute each hypothesis? Was the evidence convincing? How did the experimental design and methodology test each hypothesis? How did the data analysis illustrate the authors' points?
9. Identify how the key concepts of the reading fit into a larger assignment such as a term paper, group project, or end-of-semester presentation.
10. Write a response to the authors of the assigned manuscript. The letter should summarize the manuscript's key points; describe how the research contributes to science; laud any novel, interesting, or rigorous aspects of the research; describe the strengths in the experimental design or data analysis; and provide suggestions for improvement or future research.
11. Write a 100-word abstract for the reading. Discuss.
12. Identify the contentious issues in the reading and debate them.
13. Answer a question about the reading. Share answers with the large group and discuss.
14. Compare/contrast: Explain how this reading is similar to or different than something that was discussed previously.

competitive experiences to measure learning. If, however, our goal is to *promote* learning, instead of just measuring it, we must consider the value of group exams. Group exams are most effective with open-ended, complex questions that do not have right or wrong answers. The group process, interactions among students, and vigorous debate are intensified by an exam structure and the grade associated with it. These can be used to generate creative ideas and build the critical and logical thinking skills needed for biology. Students are willing to tackle much more difficult problems in groups than they will attempt individually (Johnson and Johnson 1975, Duren and Cherrington 1992). We find that students respond best to being required to work with a group but then to generate a written answer individually for which they receive an individual grade. This approach capitalizes on both group process and pride in individual accomplishment. To ensure that students understand and adhere to the rules, they can be asked to provide a list of their group members and to sign a "contract" that states the rules of conduct clearly. (See Figure 2.4 "Expectations.")

Table 3.2. Example of assessment that aligns with the learning goal.

Learning Goals	Learning Outcomes	Assessment
At the end of this lab, students will understand that bacteria are ubiquitous, microscopic organisms whose populations are difficult to measure.	Students will be able to explain what happens during the formation of a bacterial colony on an agar medium. Students will be able to explain why, based on the colonies that appear on the agar medium, they can arrive at only an incomplete estimate of bacterial populations in natural environments.	Predict-observe-explain: Students are given three Petri plates containing agar medium and are asked to use the materials to demonstrate that microbes are ubiquitous. First, they predict what will happen, then observe the results and explain how the results support or refute the hypothesis. *PREDICT: Touch an agar medium with your fingers and predict what you'll see in a week.* *EXPLAIN: Explain your prediction. Why do you think this will happen?* *OBSERVE: A week later, observe and record any changes.* *EXPLAIN: Did the observations match the predictions? Explain why the observations do or do not support the hypothesis. Are all of the bacteria on your fingers represented on the agar medium? Why can't you see the bacteria on your fingers but you can see them growing on agar medium?*

Set clear expectations: Rubrics

It should be clear to students how achievement of the learning goals will be measured, what outcomes (behaviors or levels of performance) are expected, and which assessments will be used for grading purposes. One way to provide students with a clear explanation of expectations is to prepare a detailed rubric. A rubric is a framework that dissects each aspect of an assignment and describes the components that are expected for various levels of achievement. Most significantly, rubrics are provided **at the time the assignment is given.** Huba and Freed (2000) claim that "rubrics explain the scoring 'rules': the criteria against which student work will be judged. More important, they make public key criteria that students can use in developing, revising, and judging their own work." They are helpful to students, instructors, and graders.

Example of a take-home exam for a biology course

Instructions: You are required to work on this section of the exam with other students in this class. List the members of your group above. After you have discussed the questions with others, compose your answer by yourself (typed). Do not write the answer as a group. Do not copy the answer from another student. Writing the answer as a group or reading or copying another student's answer is cheating and is not permitted in this class.

The *New York Times* article by Natalie Angier discusses the impact of a grain storage disease on human health (Angier 1991).

1. Explain the relationship between stored grain, fungi, aflatoxin, and liver cancer.
2. The 249th amino acid in the p53 protein is normally arginine. While researching cancer in liver cells, scientists found that DNA mutated in the 249th codon produces a p53 protein with the amino acid serine at that position. Write the changes in the DNA and mRNA that lead to this change in the protein.
3. Whether or not a chemical causes cancer can be tested in many ways. In one method, chemicals are fed directly to animals, then researchers assess whether or not the animals develop tumors. Another approach is to conduct an "Ames Test" which tests the ability of chemicals to cause mutations in bacteria. The Ames Test is based on the assumption that mutagenicity is associated with carcinogenicity, and the data it generates on mutation rates is used to predict how likely a chemical is to cause cancer. **Conduct a virtual Ames Test experiment** (http://scientificteaching.wisc.edu) to address the following hypothesis: **Aflatoxin is a mutagen.** Describe in detail the methods and results of your experiment. Discuss your conclusions.

Rubrics and students

Rubrics educate students about the standards they are expected to achieve. Rubrics can provide a clear articulation of scientific norms and language. A rubric can demonstrate conventions for writing a testable hypothesis, making clear that it is neither prediction nor theory. Students can use rubrics as a guide to evaluate their own progress and as a tool for reviewing each others' work. They can be used to evaluate individual assignments or progress toward the goals of the entire course. For individual projects, the rubric might focus on clear descriptions of what students should know, understand, and be able to do that are related specifically to that material. For evaluation of progress toward course goals, the rubric is more likely to stress overall skills and knowledge that should be gained by the end of the course.

Rubrics and instructors

Instructors can use rubrics to hold themselves to teaching according to the principles of backward design. (Refer to chapter 5 for more information.) A good rubric clearly states the learning goal and then delineates levels of achievement and how they will be judged, reminding the instructor to start with goals and work backward from there. Grading with a rubric provides information about teaching and learning achievements that can be used to guide future instructional decisions and to serve as evidence of the caliber of teaching. In addition, rubrics make transparent to colleagues the instructor's intentions, which can open lines of communication and foster discussions about teaching-related issues. Rubrics can also remove the directive voice that typically characterizes the instructions for assignments or projects. The nonauthoritative tone of rubrics sends a message that it is the student's choice to do excellent work, or intermediate work, or not to do the work at all rather than sending the message that the only correct approach is to follow the instructor's orders.

Assessment drives student learning, so rubrics should guide students to the prioritized goals. Terms should be descriptive, explaining clearly the characteristics or qualities expected for each level of performance. Similarly, descriptions should be qualitative, not laden with value judgments or comparisons. Many rubrics also include a statement about the consequences of each performance level. Terms should also define the highest caliber of performance, such as "comprehensive," "complete," "sophisticated," "exemplary," or "excellent." Regardless of which is used, each term warrants a complete description of its unique qualities. Intermediate or lesser performances deserve the same attention as exemplary performances; simply reducing the quantity of incidents is not effective or informative. Rather, a well-designed rubric articulates the qualities and consequences that characterize typical performances at each level. Consistent with scientific teaching, rubrics are iterative in nature. A good rubric is a work in progress, adjusted for context, students, and instructor goals.

Rubrics and grading

Rubrics can make grading more objective and straightforward. Many rubric categories, such as the "accuracy of information," are best shared with students because they make expectations explicit and grades less of a surprise to students. Rubrics may need to be amended for grading purposes. To be sure that graders adhere to the same standards, it may be important to include information in the grading rubric that should not be in the rubric provided to students because it is information that they should discover for themselves. For example, the specific knowledge students are intended to learn might be excluded. The effort of preparing such a detailed rubric can save time later on because grades will be consistent and appear more fair to the students.

Rubrics—an example

The examples in tables 3.3a and 3.3b are based on an activity in which students write a four-page report to propose a new genetic engineering technology to the Food and Drug Administration (FDA). One criterion, accuracy of information, is compared between the two rubrics. Most rubrics would contain other criteria as well, depending on the goals of the activity.

The first rubric does not describe the type of errors that are being assessed (factual), nor does it differentiate among errors of different magnitudes (significant vs. minor errors). In contrast, the second rubric includes both explicit descriptions about the type of qualities that typify each level of performance and brief statements about the potential consequences of that performance in relation to the assignment. For another rubric example, refer to the detailed rubric in chapter 5 for developing instructional materials, including criteria and consequences.

Tables 3.3a and b. Examples of two rubrics.

The first provides little insight into expected performance, whereas the second gives clear expectations about the work and describes the consequences therein.

Table 3.3a. *Instead of:*

Criteria	Excellent	Good	Poor	Unacceptable
Accuracy of information	No errors were made.	1–3 errors were made.	4–6 errors were made.	More than 6 errors were made.

Table 3.3b. *Try:*

Criteria	Level of performance			
	Sophisticated	Good	Needs improvement	Unacceptable
Accuracy of information	No factual errors were made. Your work will be very useful in aiding the reader make a decision about whether this genetic engineering technology would be a significant contribution as an alternative method to pesticide use in agriculture.	No significant errors were made. The reader recognizes any errors as the result of hasty conclusions or oversights. Your work is usable for making decisions about employing this technology, but would be considered more reliable if you were more careful in proofreading your work.	Enough errors were made to distract the reader, but the reader is able to use the information to make judgments. The technology will appear more useful if the reader is able to decide what evidence is reliable.	Your proposed technology is highly improbable because there are so many factual errors. The reader cannot depend on this report as a source of accurate information, or you have included so little information that the reader is not sure what the technology is about. It will not be approved by the FDA.

The second rubric is adapted from Huba and Freed (2000).

Closing remarks

Assessment is probably the most ignored part of good teaching, so it is often the most intimidating to instructors. It need not be onerous if it is integrated into teaching in natural ways that include both large and small assessment events. Once instructors become familiar with the iterative process of assessing, teaching, and reassessing, and of teaching students to evaluate themselves, teaching without ample assessment feels like groping in the dark and it's hard to imagine teaching without it.

CHAPTER 4

Diversity

Diversity Defined

Human diversity refers to the variation of human experience, ability, and characteristics. Diversity should be considered in teaching because (1) we owe all students education about the diverse world they live in, (2) diversity enhances learning, and (3) each student will experience the classroom differently from everyone else in the class. Differences in education, experience, cognitive styles, personalities, abilities, cultural backgrounds, physiology, and innate characteristics conspire to make the classroom experience unique for each student. Incorporating human diversity into science education and recognizing student differences and our reactions to them can prepare students more effectively for the global community, enable us to reach more students, and enhance the vibrancy and quality of research and teaching on college and university campuses (Nelson and Pellet 1997, Turner 2000, Milem 2001, Antonio 2002).

Why Diversity?

Diversity enhances science

A scientific community derives its health and vigor from diversity. Good science necessitates teamwork typified by members with different approaches, experience, and ways of thinking. Evidence from controlled research studies shows that heterogeneous groups are more creative in problem solving than homogenous groups (Cox 1993, McLeod *et al.* 1996). Groups with a minority view defend their solutions to problems more effectively than those that do not contain a minority view (Nemeth 1985, Nemeth 1995). In studies of mock juries, those that contained members of ethnic minority groups deliberated more effectively and processed information more carefully than juries that lacked ethnic diversity (Sommers 2006). Studies of real life reinforce the results from controlled studies. A study of the teams that produced hit Broadway plays and high-impact scientific papers found that diversity of experience was a common feature of the most effective groups—the mixture of novices and veterans produced the most creative and successful teams (Guimerà *et al.* 2005). In short, fostering diversity is good for the creative process. Innovation and collaboration are hallmarks of science. Therefore, advancement in science necessitates diversity.

Diversity enhances education

Studies of students in many different colleges and universities show that a diverse student body produces better-educated graduates with more highly developed cognitive abilities, interpersonal skills, and leadership abilities (Astin 1993, Gurin 1999 and 2002). Graduates will enter a global community in which they will confront a diverse workforce and a range of human experiences. Their education should prepare them for their future by offering them opportunities to learn about a world and people beyond their own experience and to work effectively with people whose backgrounds differ from their own (National Research Council 2003). Therefore our students, our universities, and science itself benefit from building a diverse scientific community and presenting science in a global context.

Diversity and global competitiveness

Concern about the global competitiveness of the United States in science and technology is mounting (National Research Council 2006). A recent blue-ribbon panel that studied this issue recommended strengthening education in science and technology to retain more top-quality students in these fields (National Research Council 2006). Therefore, we need to take full advantage of the diversity of brainpower that we train. In many fields, such as biology and chemistry, nearly 50% of the Ph.D.s have been granted to women for years, but the proportion of women in the faculty does not reflect this. According to 2002 statistics about the "top 50" biology departments in the U.S. (defined by amount of NSF funding; Nelson 2004), only 15% of full professors on biology faculty are women. Since women are well represented among the very best graduate students, by not utilizing these gifted women in academic science, the professoriate is not drawing on some of the best talent.

The picture for minority groups is bleak. Among full professors in the top 50 biology departments, 11% are African-American, Hispanic, Asian, or American Indian/Alaska Native (Nelson 2004), yet these races combined comprised 25% of the overall American population in 2000 (U.S. Census 2000). These statistics, however, mask the far lower representation of certain groups. For example, there is one Native American full, two associate, and one assistant professor among this group. The representation of black, Hispanic, and Native American people across all levels of faculty in these departments hovers around 1%, and the only minority group that is better represented is Asians, who constitute 8% of the total faculty.

Diversity in the college classroom

So why is it the responsibility of college and university instructors to consider diversity in their teaching? As many reports have shown, the diversity problem is partly rooted in the pipeline of people delivered from each educational level to the next. Given the intellectual, educational, scientific, social, and economic need for diversity among our scientists and consequently among our science students, it is incumbent upon those in higher education to examine our track record and understand our contribution to the problem. Studies addressing the attrition of students from science majors have shown that many of those lost are the top students in introductory courses and many of those who leave the sciences go on to graduate work in the humanities, indicating that there is no obvious academic reason for their departure. Many of these students (particularly women) leave certain fields because they find the world of science alienating or even hostile (Seymour and Hewitt 1997, National Research Council 2003). Many others leave because they find science in introductory science courses to be about memorization, not about problem solving and curiosity (Tobias 1990). In other words, science courses are not weeding out those who don't belong in science; they are driving away students who seek the intellectual substance of science but are not finding it because the introductory science courses misrepresent the nature of science.

Statistics from the National Science Foundation illustrate these trends starkly (Table 4.1).

Some fields have aggressively embraced the challenge of diversity. Engineering has a strong focus on preparing students to function in a global economy, which necessarily means addressing diversity in the classroom (National

Table 4.1. Attrition of Students from Science and Engineering Majors in College.

	Total S&E	Bio/Ag	Computer	Engineering	Math/ Stats	Physical	Social/Beh
White	-1.2%	-17.7%	37.8%	-44.8%	31.4%	-25.4%	70.0%
Female	9.5%	-25.0%	169.9%	-41.7%	29.7%	-31.3%	56.2%
Male	-7.8%	-11.2%	30.6%	-43.8%	58.0%	-12.7%	105.3%
Asian/Pacific Islander	6.9%	-11.4%	139.4%	-35.4%	142.7%	-9.8%	139.0%
Female	16.9%	-17.4%	312.9%	-35.3%	110.2%	-5.8%	104.5%
Male	1.4%	-2.9%	105.1%	-32.7%	134.9%	-3.0%	218.0%
Black	-10.8%	-32.4%	15.0%	-54.9%	51.5%	-10.3%	52.0%
Female	-9.1%	-38.5%	49.0%	-54.4%	28.4%	-4.5%	34.7%
Male	-3.8%	3.8%	-23.1%	-14.4%	86.9%	0.1%	97.0%
Hispanic	-8.8%	-30.6%	-41.2%	-48.1%	26.3%	-12.2%	49.0%
Female	-4.8%	-35.9%	-53.4%	-39.3%	7.9%	7.0%	30.6%
Male	-17.6%	-43.4%	-1.9%	-67.8%	63.6%	-29.6%	89.2%
American Indian	-1.6%	-9.1%	73.8%	-53.8%	47.2%	-40.6%	52.2%
Female	0.5%	-30.1%	203.4%	-37.8%	123.7%	-47.0%	30.1%
Male	2.8%	32.2%	37.0%	-56.0%	32.5%	-35.6%	104.8%

National Science Board 2002 Statistics on Loss and Gain of Undergraduates in Science and Engineering Majors. The numbers represent the difference between the number of students who graduate in each area divided by the number who declared as freshmen an intention to major in the area, expressed as a percent. Highlighted boxes indicate the most substantial loss of students from the major from first year to graduation.

Research Council 2003). This agenda is driven in part by the attention paid to the issue in industry. Sun Systems, for example, has a program on diversity and engineering in a global environment that illustrates the necessity of confronting the issue in order to be competitive (Sun Microsystems 2006). Many engineering programs offer courses and experiences that teach students aspects of global diversity and inclusivity that will affect their abilities as engineers (these include MIT, Harvard, and Cornell). The biological and physical sciences do not seem to have as systematic and universal focus on diversity and globalization.

Imagine a classroom with no diversity—a room full of people of uniform race, gender, ethnicity, physical abilities, intelligence, experience, attitudes, religion, sexual orientation, habits, dress, height, shoe size, learning styles, and political outlook. What would be lost? What would you miss? What would you do to increase the diversity in the room?

Diversity of Students

Diversity of learning styles

Most manifestations of diversity are invisible and unknown even to the person who harbors them, but these elements of diversity can have profound effects on learning. Childhood experiences, education, relationships, and personality all contribute to the development of the mind, making thinking one of the most idiosyncratic of human activities. In a single classroom, there may well be students whose thinking is dominated either by intuition, logical deduction, or inductive reasoning. Some people favor learning details and deriving a global view from the facts and pieces. Others may learn best by starting with a principle and supporting it with examples. Some are driven by curiosity, others by emotion; some are visual learners, others are auditory, and still others are kinesthetic learners, who learn best by doing. Some learn best in a competitive context; some are paralyzed by competition. And some may prefer to approach learning differently in different situations.

The field of learning styles offers some insight into the complex landscape of ways that students approach learning. The 71 distinct models and five families of learning styles identified by Coffield *et al.* (2004) indicate the convoluted nature of the field. To make it even more difficult, there is little evidence to suggest that using learning styles as a basis for teaching methods enhances learning, so the scientific teacher has little to utilize from the field. However, many of the models showed promise, and, with further evaluation, will likely provide important guidance to learning and instruction. For now, we just remind instructors that there are likely to be as many learning styles as there are students in the classroom; thus it is essential to choose a diversity of effective teaching methods to engage students in constructing knowledge.

Diversity in race and gender

Variations in learning styles among students may be hard to detect, but other differences among people are easy to see. Race and gender are among the first attributes we notice about people. Moreover, most people identify much more strongly with the group defined by their gender or race than the group defined by their learning style, and society makes much sharper judgments based on race and gender. The impact of race and gender on the classroom is evident and important, so it is worth spending some time thinking about the research in this area. Conscious thought about race and gender in the classroom can inform teaching methods and help avoid inadvertent discriminatory behaviors. Science educators also have a responsibility to expose students to human experiences beyond their own. Therefore, choosing examples of biology that are drawn from diverse settings or involve diverse people is important in broadening the learning experience in our classrooms.

Incidentally, girls can do math and science! Much has been made in the popular press about ostensible genetic differences in aptitude for science between men and women. Careful analysis of the research on this issue reveals that although the field is highly controversial and much is made of data on both sides of the argument, there is little support for the assertion that men are better able to do math and science. The assertions about boys' superiority in math, in particular, are made sufficiently often that it is worth examining the research. One highly publicized result is that math SAT scores have been slightly higher for boys than girls for years (Spelke 2005). However, many other lines of evidence indicate that boys and girls perform similarly in math (Hyde 2005). Boys' math performance is either slightly better or the same as girls' in early childhood (Hyde *et al.* 1990) and by high school, the gap has closed and girls do as well and in some studies appear to do *better* than boys (Hyde *et al.* 1990, Hyde 2005). Levine *et al.* (2005) found that spatial ability, which can affect success on standardized math and science tests, may be more influenced by socioeconomic class than by sex. The authors' findings were explained in part by the observation that boys in economically advantaged families are more likely to play with toys that build spatial skills, such as video games, which contribute to success in certain types of math problems. This last result underscores the point that even if there are measurable differences in performance—which is questionable based on the available data—math aptitude is influenced by engagement in certain activities, and spatial skills are learned, not entirely innate. By the time they reach college, there is no detectable difference in math performance of men and women. A recent study showed that women college students take as many math courses and perform better in them than their male counterparts (Evans *et al.* 2002). These results show that the SAT scores, which are poor predictors of college performance generally, underpredict women's math performance in college. The role of learning and culture in math ability is also highlighted by the large differences in math ability among children in different countries. Girls in Taiwan and Japan, for example, perform far better than boys in the United States (Lummis and Stevenson 1990).

Other cognitive differences between men and women have been studied as well. Girls consistently outperform boys in tests for reading and writing ability (National Center for Education Statistics 2000, Weiss *et al.* 2003) and this gap appears in international studies as well as those conducted in the United States and over a large number of studies spanning two decades (Hyde and Linn 1988, National Center for Education Statistics 2000).

Socialization is likely to be more important than differences between men's and women's performance on math tests. One of the most significant influences is stereotype threat. This is the phenomenon that when people are reminded of their gender or race, their performance on certain tasks is more likely to conform to a stereotype. For example, when Asian women were reminded of their race, they performed better on math tests and when reminded of their gender, they performed worse (Shih *et al.* 1999). Women's math performance diminishes when men are present during the test-taking (Inzlicht and Ben-Zeev 2000, Quinn and Spencer 2001). When women are told that the test indicates gender differences, they will perform less well than men, but in contrast, they will perform just as well as men if they are told that the tests show no gender differences or that the tests are not diagnostic of ability (Spencer *et al.* 1999, Davies *et al.* 2002).

The aggregate of this research indicates that by the time students reach the college classroom, men and women are likely to perform similarly in math, and women are likely to perform somewhat better in endeavors involving verbal skills. Differences are more likely to be associated with social factors than innate or learned behavior, and expectations and attitudes toward gender and ability affect performance.

So what does this mean for science? We don't know much about the combination of skills that is needed to be an effective scientist. Most of us would probably guess that global intelligence, math skills, verbal skills, social skills, and hard work are all needed to be successful in scientific research. But how much of each is required? That's the impossible question. It is impossible to answer partly because these characteristics are impossible to measure accurately, they change over a lifetime due to experience and learning, and so many different types of scientists are successful. This last point is probably the most important—diverse approaches to science are successful, and diversity keeps science vibrant and dynamic, and so there is no recipe of features that produces a great scientist. However, the research illuminates the importance of having high expectations of all students and not activating stereotype threats, which diminishes performance. In our classroom role, we should treat our students as individuals and not judge their abilities or potential based on a group characteristic, real or imagined. As will become apparent, however, dismissing our unconscious biases about race and gender is harder than most of us think.

Diversity and Unconscious Bias

Although most people consider themselves fair, objective, and unprejudiced, **everyone** brings bias and prejudice to interactions with other people. Copious research, which includes both controlled experiments and analyses of real-life situations, shows that we **all** apply assumptions and biases to our judgments of people and their work. For example, if subjects are asked to judge a set of credentials, they rated the quality of the materials lower if told the credentials were for a woman than for a man (Olian *et al.* 1988, Steinpreis *et al.* 1999). When asked to rate a subject's verbal ability based on a paragraph of text, evaluators gave a lower rating if they were told the paragraph was written by an African-American person or a woman than if they thought it was written by a white person or a man (Biernat and Manis 1994). One research group sent the resumé of a real person to a large group of academic psychologists and asked them whether they would hire the person. The frequency of positive answers was substantially higher if the resumé bore a man's name than if it bore a woman's name (Steinpreis *et al.* 1999). In another study, evaluators were given a description of a person's career path and photograph of the subject and asked to judge whether the person's success was due primarily to ability, luck, or political skills. If the subject was attractive, the attribution was quite different than if the subject was unattractive, but the trend was gender specific: more raters attributed success to ability for attractive than for unattractive men, but the response was reversed for women subjects, whose success was twice as likely to be attributed to ability if they were *unattractive* (Deaux and Emswiller 1974, Heilman 1985).

Similar observations have been made in studies of real-life evaluations. For example, an analysis of postdoctoral fellowship applications in Sweden compared the correlation between "impact scores," which were based on the number and impact of the applicants' publications, and "competence scores," which were used to rank the applicants and award fellowships. For the male applicants the impact scores were tightly correlated with competence scores. For women applicants, however, impact scores were much less well correlated with competence ratings (Figure 4.1). This means that women needed to have many more publications (the equivalent of 3 more papers in *Science* or *Nature* and 20 more

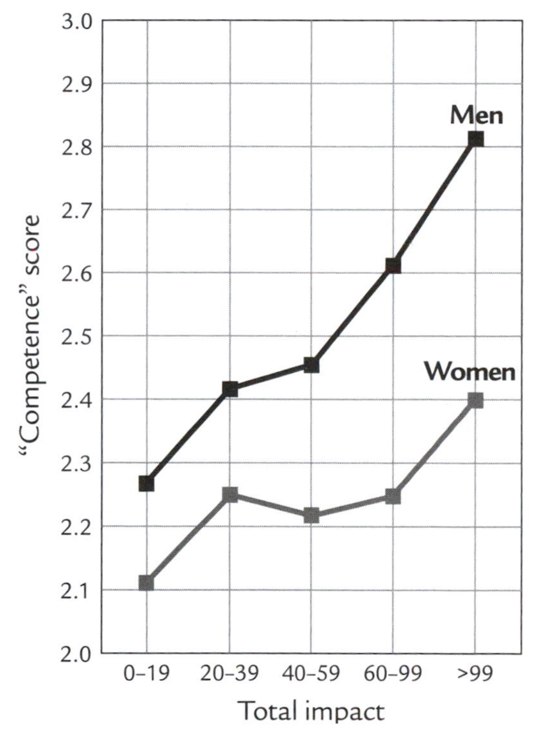

Figure 4.1 Relationship between total impact (number of papers x impact factor of journals) and overall competence rating (determinant of whether a fellowship was awarded) for applications for postdoctoral fellowships in Sweden. (Wenneras and Wold 1997, used with permission)

papers in a specialty journal such as *Neuroscience*) in order to achieve the same competence rating as their male counterparts (Wenneras and Wold 1997). In professional music, symphony orchestras experienced a 60% increase in selection of women following the installation of screens that obscured from the evaluators the gender of people auditioning (Goldin and Rouse 2000). These research studies illustrate the power of our unconscious biases. Our rational evaluations of quality are shrouded in a fog of cultural expectations of people with certain characteristics.

A key element of all of these studies is that in none of them does the gender of the evaluator show a significant effect on the results. This means that *both men and women are likely to bring similar unconscious biases to the classroom*, even if we are the subject of those very same negative biases.

Unconscious biases and interactions with diverse students

It is easy to imagine how our expectations of students might be shaped by unconscious bias. When we interact with a female student, do we assume that she is less competent in science than a male student? When we grade an African-American student's paper, are we harsher on the writing because of an unconscious expectation of lesser writing skills? Students live up or down to expectations, so we do them a great disservice by expecting less than top ability and performance from every student.

More research is needed about diversity in college science education

One frustration in considering diversity is that there is only a small literature about the types of interventions that have a positive effect on students with different learning styles, race, gender, or other differences. Much research is needed to assess the value of the impact of inclusive teaching strategies in college science courses for women and minorities (Rosser 1991). The NIH has been working to improve the percentages of minorities in biomedical research for over 30 years. In a 2005 report, the National Research Council reviewed the NIH's evidence of success in *Assessment of NIH Minority Research and Training Programs*. The major finding of the report was that the NIH did not provide sufficient quantitative evidence of success, including basic information such as the percentage of minorities who have participated in the programs or who have chosen biomedical careers. If these programs are to continue to receive funding and be broadly adopted, demonstrable success is imperative.

The "prejudice paradox" and the nature of bigotry

The reader may note an inherent paradox in this section. First, we present information about learning styles and race, ethnicity, and gender that suggests that educators may want to consider diversity in choosing teaching methods and content in order to optimize learning for diverse students. Then we detail the insidious power of unfounded, unconscious biases about women and minorities. Is there a "prejudice paradox"? Are we advocating using research that makes generalizations about groups to shape our teaching and simultaneously cautioning against making assumptions about people because of their group affiliation?

Both bodies of work can be accommodated by a simple cautionary thought: people differ in many ways—some that we can see or measure, some that are unknown or unknowable, and some that are imaginary, but assumed. We should treat every student as an individual who has the potential to be outstanding in science, but we should also assume that there is tremendous diversity in ability, experience, cognitive style, social preferences, and myriad other characteristics among our students. Efforts to accommodate this diversity can be informed by the characteristics associated more often with women, Native American, or African-American learners, but we should never assume that any individual has characteristics associated with a group. After all, that is the definition of bigotry.

How do people react to evidence about discrimination?

Many academic scientists believe that we have solved the problem of discrimination, saying that we judge people based on quality alone, that we treat all our students the same, and that discrimination is a thing of the past. A recent study at the University of Wisconsin (Pribbenow *et al.* unpublished) showed that department chairs (mostly white men) rated the climate for women and minorities more positively than did women and minorities. When confronted with data that demonstrate discrimination, many people quickly generate hypotheses about how the data are flawed and do not really show that discrimination is alive and well.

The Chancellor of the University of Wisconsin-Madison, John Wiley, remarks that nothing elicits rationalizations as quickly as gender issues. In Wiley's words:

Here's how it usually goes:

I comment that women students consistently have significantly higher GPAs than men students, here and around the country, and that this has been a stable fact for many years. Women also graduate faster (shorter time-to-degree), and in greater proportion (higher graduation rates).

The immediate "explanation" offered by men for this uncomfortable piece of data is, "Yes, but that's because women enroll disproportionately in 'easy' and non-quantitative subjects."

I then point out that women students have significantly higher GPAs in EVERY major, including all the sciences, mathematics, and engineering, so their explanation doesn't hold water.

To this, the response is "Well, but only the 'best' women major in those fields, whereas more 'average' men choose those majors, so you are comparing the very best women with average men."

But, when we control for academic preparation, comparing only men and women who had similar high school GPAs, class ranks, and test sores, the women still get higher college GPAs in every major. When I point that out, the next response is something like: "Well, women are less self-confident and overcompensate by studying harder. Men lead more balanced lives and are less grade-obsessed."

And so it goes. For every new fact there is a new or revised theory or explanation— often presented with great confidence. I am constantly astounded at the ability and propensity of academics to manufacture instant explanations for any morsel of data without demanding other relevant data or context or control information. I only hope they are more meticulous in their own research!

John D. Wiley
Chancellor and
Professor of Electrical Engineering
University of Wisconsin-Madison

Why are many people resistant to the idea that bias exists and that discrimination is quite alive in our society? No one knows the answer, but there are some speculations. Most people would like to believe that they live in a fair world, and these data challenge that belief. Many people see women and minorities in their universities and assume, therefore, that things are going well in the universities' efforts to diversify. Even more fundamental is the flip side of discrimination: if one accepts that others suffer from discrimination, then those who do not suffer discrimination necessarily benefit from the privilege of being white or male (McIntosh 1990). Few people want to believe that their own success was due in part to how they look. Scientists are particularly uncomfortable with this idea because they cling to the image of the scientist as a lone ranger, a pioneering cowboy striking out where no one has gone before, alone and with nothing but his ingenuity and ability to deal with adversity as his aids (Lawrence 2006). Perhaps some scientists believe that their successes are diminished if they did not arise from hardship, deprivation, and a tough fight, but rather were assisted and protected by the privilege of being white and male and therefore part of a social group that has automatic credibility and is connected by a network of members of the same group. This is simply a hypothesis, but it is consistent with extensive study of the privileges enjoyed by whites and men that are largely unacknowledged (Kulis *et al.* 1999, Sagaria 2002).

Every educator is responsible for providing equal opportunity in education. No one can eliminate his or her own biases and prejudices, but we all have the obligation to minimize their impact on our treatment of students. It is natural to bring what we know about our students to bear on how we teach them—we teach first-year students differently from seniors, majors differently from non-majors—but we need to confront our assumptions about individuals and give every student the chance to excel. The result will be a more inclusive scientific workforce, which will strengthen the practice of science.

In conclusion

In short, science and education will be strengthened by attracting diverse intellects and personalities to our classrooms, providing environments in which diverse people succeed flourish, and by evaluating their long-term successes. We have compiled examples from many other disciplines to contribute to our understanding of diversity, but we are a long way from fully understanding the complex effect of diversity in the science classroom.

The preceding section will offend or discourage some readers. That's okay. This chapter is intended to lift the reader out of his or her comfort zone and make him or her consider these emotionally charged and intellectually challenging issues. Teachers who grapple with the complexity of social science data about students and the messiness of learning styles, race, and gender issues in the classroom are better equipped to reach diverse students than those who avoid the issues. Each will arrive at different conclusions from evaluating the data, and every teacher develops personal strategies for reaching many different kinds of students. That's the power of having a diversity of educators in our university classrooms.

Diversity in Practice

A teacher's challenge is to treat students as individuals, without expectations shaped by assumptions based on a group affiliation. Even more challenging is to maintain high expectations for, and confidence in, every student, even those who have performed poorly. Assuming that a student isn't fit for science based on one low exam grade may be as damaging and unfair as assuming that the student isn't fit for science because she is a woman.

People of certain races, ethnicity, or gender may have social or intellectual preferences that shape instructors' choices of teaching methods. For example, some research shows that Native Americans learn better, on average, in a cooperative group than in a competitive system that spotlights the achievements of one student (Swisher 1990). Connections of course material to applications to enhance people's lives enhance many students' learning, but have a more pronounced effect on women (Rosser 1991). These studies—and many more—present generalizations that do not apply to all members of any group, but illustrate that some of our students may have learning styles that differ from each other's or from their instructor's. Teaching methods need to accommodate a range of styles, preferences, and experiences.

In this section, we provide suggestions for integrating diversity into science courses. Some of these models have been rigorously tested in classrooms, some have been used in science, some we have tried, but all have some merit based on literature or the authors' experience. Activities that encourage active learning and assess student learning can engage diverse students. We leave it to our readers to determine which elements are most useful for them, and we hope to foster an intellectual dialogue about diversity in the classroom.

Try some strategies to create a more inclusive classroom

There are many ways to integrate diversity into science education. Table 4.2 provides a quick overview of some strategies to make classrooms more inclusive based on the themes of this chapter.

Reduce the impact of bias

Instructors can act as their own monitors. Being aware of the potential for bias is the most effective guard against it. Are students being treated based on their behavior and performance, or on expectations that have nothing to do with them?

Table 4.2 Suggestions for converting a noninclusive classroom to an inclusive classroom.

Noninclusive Classroom	Inclusive Classroom
Scientists in the examples are all white men.	Examples include important contributors to science who represent various ethnicities, races, and genders.
Many of the analogies and metaphors used are sports-, military-, or construction-related.	A broader range of analogies is used.
Class is held on the second floor of a building with no elevators.	Class is held in a wheelchair-accessible room.
All lectures use text-rich PowerPoint; all homework involves reading long passages with few visuals; all study guides involve reading questions and answering in essay format.	Lectures include a mixture of visual representation of data, images, or pictures; include opportunities for students to manipulate information through concept maps or group discussion.
Underrepresented minority students are all grouped together or completely separated.	At least two women or members of ethnic minorities are included in each group.
Homework and exams are completed individually and graded on a curve.	Some activities depend on the contribution of everyone in the group.
Instructor's questions always have a "correct" answer.	Answers to questions have many correct answers, require consensus of the group, or require the class's collective knowledge to be answered.
Exams are timed for 50 minutes.	Instructor gives some take-home exams and students have a week to complete them.
Instructor doesn't use a microphone in a large lecture hall.	Instructor ensures that everyone can hear by using a microphone.
Instructor uses one type of teaching method (lecture) and exam (multiple choice).	Instructor integrates a variety of active learning exercises and assessment tools regularly into class and homework. Exams include multiple formats.
Ecology section of introductory biology is relegated to the end of the semester or is deemed "simple."	Ecology section is treated with equal value and scientific rigor as molecular biology sections.
Grading is done without discussion or a rubric.	Grading schemes are discussed and done according to a rubric.

When an unfamiliar student approaches with a question, is the level or depth of the instructor's answer shaped by their sex, race, age, or manner of dress? Do the thought experiment: if a student clanks toward me in black leather and chains, with a shaven head and pieced body parts, would I answer his questions differently than if he dressed the way I dress? How might this student perceive *me*?

Enhance all students' exposure to the diversity of human experience

Diverse examples should be used throughout teaching, regardless of the subject. Images of diverse natural habitats, races of people, and types of human activities can enhance the learning experience. A unit about the impact of human activity on biodiversity, for instance, can be illustrated with examples from other parts of the world or in which human issues that our students do not normally confront interface with ecological issues (the effect of poverty in developing countries on decisions about protection of the rain forest, for example). Similarly, a discussion about pathology can include diseases that affect people across the globe, introducing our students to experiences beyond their own. Images of people should include images of men, women, and people of all races and abilities.

Make classrooms inclusive and inspiring for women and people of color

A discussion of the contributions and showing images of diverse scientists sends a subtle message that everyone is welcome in the classroom and that everyone can be successful. One study showed that people evaluated African-Americans more favorably for about a day after seeing images of great African-Americans, such as Martin Luther King, Jr., than if they were shown no images or negative images of African-Americans (Blair *et al.* 2001, Dasgupta and Greenwald 2001). Extrapolating from these results, we can perhaps reduce bias toward women and minorities by providing our students with more examples of successful women and minority scientists.

Reduce the effect of personal bias in grading

When criteria for grading are established **before** the grading is done, expectations will be clearer and the effect of personal bias will be lessened. One study showed that people construct criteria after the fact to justify unconscious discrimination, and having clear criteria for evaluation reduced their bias (Uhlmann and Cohen 2005).

Teach students about bias and diversity

Diversity exercises in the classroom can help students understand the unconscious biases they bring to the classroom and make clear that each person brings unique strengths. Have them examine behaviors (who talks and how much, who does the work, whose ideas prevail) when they work in small groups and discuss what they observe. Consider assigning reading about diversity and discussing it. Even if this material is not directly relevant to your scientific content, it **is** relevant to your students' ability to learn in your class. You also owe your students from majority groups some exposure to diversity issues before they enter the workforce where they are guaranteed to confront them. Many resources are available to foster these discussions, including the Diversity Institute created by the Center for the Integration of Research, Teaching, and Learning (CIRTL 2006).

Use learning styles to increase self-awareness

Let students know that there are many classifications available and that they may fall anywhere on the spectrum of learning styles. Certain types of classroom experiences will make perfect sense to them, and they'll be able to do their best easily; other experiences will be a struggle. It is important for them to understand that their performance may be influenced by any number of factors: it may be culturally mediated; it could be a function of their personality; it might be affected by the classroom environment, social cues, or physiology. A little bit of knowledge helps students develop awareness that not everyone—including the instructor—approaches learning in the same way, which can generate dialogue between students and instructors about learning. In addition, it may be important for students to learn that even the researchers don't have a consistent message about which approaches are valid.

Beware of labeling students

The appeal of categorization is seductive. It's enticing to think that a simple classification of students will make the classroom operate more smoothly and help everyone learn better. Most students, however, do not fit neatly into the learning styles categories. In addition, some researchers warn that any classification is antithetical to creating an inclusive classroom because it may promote the very biases that the instructor is trying to erase (Reynolds 1997). In contrast, other researchers claim that when students know their learning styles, diversity is the norm because no one is deemed "special" or "disabled" (Dunn and Dunn 1992). Others consider learning-style classifications to be useful simply because they can encourage students to expand their repertoire of preferred learning styles and be better trained for the new workforce, or because they can be used to enhance—rather than diagnose—learning when used as tools for self-development (Garner 2000). Still others claim that knowing about learning styles has little effect on students' self-concept and therefore minimal effect on the quality of learning (Desmedt *et al.* 2003).

Approach learning as a scientist

Because the evidence that addressing different learning styles promotes higher student achievement is marginal at best, evaluation is critical. Research results—both from the literature and from practice—should guide instructional decisions. Try a diversity of techniques and evaluate how well they work for both students and instructors in the classroom. Do the interventions improve student achievement or motivation? Do they foster a more dynamic, egalitarian learning community? In addition, while it's no more likely that an instructor can be pigeon-holed into a learning style category than that a student can be, certain approaches will resonate with each person. Learning style models give instructors a way to gauge how diverse their own teaching methods are, what type of approaches they prefer to use, or identify new teaching methods that expand their usual repertoire.

Contribute to the diversity of the scientific community

By creating a classroom culture in which people of diverse cognitive styles, social styles, experience, race, ethnicity, and gender thrive and feel included, college educators can enhance the intellectual vigor of their classrooms and contribute to building a pool of confident, ambitious graduate students. The confidence and sense of belonging that these students acquire as undergraduates will buoy them through difficult times when they feel disenfranchised by the scientific community because they display characteristics that mark them as different.

CHAPTER 5

A Framework for Constructing a Teachable Unit

Teachable Unit Defined

A **"teachable unit" is comprised of instructional materials** that are based on scientific teaching principles and meet the following criteria:

1. It is designed to engage all students in learning.
2. It is designed to provide feedback to instructors and students about learning.
3. It explains how the activities and assessments are designed to help diverse students achieve the learning goals.
4. It contains learning goals that represent that nature of science.
5. It contains all the information that another instructor would need to teach the unit and accomplish the learning goals, including student materials and handouts.

A teachable unit can contain materials for one class, a series of classes, a unit, or an entire course. (Smaller units, such as a single activity, are considered "teachable tidbits.") The steps outlined in this chapter put scientific teaching principles into practice by integrating active learning, assessment, and diversity into the development of a teachable unit.

Why a Teachable Unit?

A teachable unit puts scientific teaching principles to work

A teachable unit is more than student instructions and lecture notes. It also explains the rationale for the design of the unit. It clarifies what students are supposed to be learning, how well they should understand it, and that the responsibility for learning is theirs. Learning goals are paramount, and it is clear *how* each activity and assessment measures the intended learning outcomes and helps a diversity of students achieve those goals. Framed this way, the unit not only provides the instructor with an outline that keeps the unit on track and on time in class, it also helps the instructor make teaching choices that are based on priorities that have already been established.

A teachable unit makes instruction transparent

A teachable unit explains what students will learn, and it explains how they will learn it. Because the reasons for instructional decisions are clear, students better understand what they are supposed be doing, why they should be doing each activity, and how the activities help them achieve a specific learning goal. Therefore, they can develop and implement ways to self-monitor progress toward the learning goals. In addition, a teachable unit encourages sharing of ideas and peer review of teaching among colleagues because other instructors can see the logic and rationale behind the unit.

A teachable unit reinforces scientific teaching

The format of a teachable unit provides a mechanism for collecting data about teaching for publication or for teaching reviews. Clearly defined learning goals and assessment tools serve as the framework for data collection. The teaching experience provides an opportunity to test the hypothesis that the unit includes effective teaching methods that are designed to foster learning.

A teachable unit focuses on learning

Four key elements comprise a teachable unit. Each of these components is described in detail in the next section, "Building a Teachable Unit" and are listed in Table 5.1.

Table 5.1 Framework for a Teachable Unit

Title
Learning Goals and Intended Outcomes: What will students know, understand, and be able to do? What performances or behaviors will indicate achievement of the goals?
Assessment: How will students and instructors gauge student learning throughout the unit?
Activities: What will students and instructors do to engage a diversity of students in achieving the goals? Briefly describe the schedule of events.
Scientific Teaching How does the teachable unit address scientific teaching with respect to the following themes? How does each component align with the learning goals? **Diversity** **Active learning** **Assessment**

Table. 5.2 Review Rubric for Teachable Units

Criteria	Levels of Completion		
	Comprehensive	Intermediate	Cursory/Absent
Learning Goals and Outcomes Goals—What students will know, understand, or be able to do Outcomes—What performances or behaviors will indicate accomplishment of the goals	**Overall: Students will experience the nature of science in this unit and know what is expected of them.** It is clear what students will know, understand, and be able to do after they have completed this unit. The goals are challenging, interesting, and appropriate for the intended students. It is clear what knowledge (concepts, topics, theories, facts, and terminology) students are expected to learn and what behaviors and performances typify understanding.	The goals are clear, but they do not entirely represent the nature of science. Students need more information to know what is expected of them. The prior knowledge that students are expected to have is somewhat inaccurate; this unit may therefore be too challenging or simple.	The goals do not represent the nature of science, or they are otherwise inadequate. For example, they may be too vague, ambiguous, broad, ambitious, detailed, or focused. There are no descriptions of expected student performances or behaviors; students will not know what is expected of them.
Active Learning How students will engage actively in learning	**Overall: Students will be actively engaged in learning.** The activities follow a logical progression within a unit and will effectively engage students. It is clear how the activities elicit students' prior knowledge and address common misconceptions.	Activities should be more clearly tied to the learning goals, interesting, or student-centered. There are not enough student-centered activities in this unit to elicit prior knowledge, to construct new knowledge, or address common misconceptions.	Activities are exclusively teacher-centered. It is not clear how the activities will engage students in learning, how they build on students' prior knowledge, or how they address misconceptions.
Assessment How instructors will measure student learning How students will self-assess learning	**Overall: Students will know what is expected of them and will receive regular feedback about learning.** The assessments provide instructors and students with useful feedback about student learning throughout the unit and at the end of the unit. Assessments are designed to drive student learning toward the goals. Criteria for evaluation	The assessments will measure student learning at some key points, but students will need more feedback during the unit. Alternatively, students will need more opportunities to evaluate their own learning during the unit. Rubrics could provide more clear descriptions of the expected performances and the consequences of each performance.	There are no formative assessments in this unit. The assessments do not adequately measure progress toward or achievement of the learning goals. The assessments do not provide useful feedback to students about learning.

How all students will be included in learning	**learning.** It is clear that the unit is designed to enable students to construct their own learning in the context of their own minds and to engage diverse students. It is clear that the unit addresses multiple aspects of student diversity, such as race, gender, and abilities. Varied teaching methods are used to address different learning goals and engage a diversity of students. A diversity of content, examples, or metaphors is used that are not offensive.	dents to construct their own learning, but could include more diverse or effective teaching methods. The unit could use more examples that reflect student diversity in cultural background, gender, learning skills, or physical abilities. The unit includes diverse teaching methods, but otherwise does not address student diversity.	foster student responsibility for learning. The unit does not address diversity or includes potentially off-putting examples.
Alignment How the unit aligns with learning goals	**Overall: The unit's activities will help students achieve the learning goals. The assessments will measure student achievement of the goals and provide regular feedback about learning.** It is clear how the unit's activities and assessments are aligned with the goals. It is clear how the unit addresses the themes of diversity, assessment, and active learning.	In some cases, it is not clear how the activities help students meet the learning goals, or how the assessments provide feedback about the goals. Certain goals are over- or under-assessed through activities or assessments, which will give students the wrong idea about what is important to learn. There are too many activities for the time period; students will likely feel this unit is "busywork."	The activities and assessments do not align with the learning goals. Students will be confused about which learning goals are important.
Teaching Plan What the instructor and students will do	**Overall: Instructors will understand the schedule of events.** The plan includes a clear schedule of events for activities and assessments for both the instructor and the students. The sequence of events is logical and aligned with goals. Detailed instructions are provided so that another instructor could easily implement the unit, including guiding questions, tips, and supporting materials. Detailed instructions are provided for students.	The schedule of activities is described broadly, but more detailed instructions are needed for another instructor to implement. The order of events is somewhat logical, but more information is needed to be useful to other instructors. Some minor factual information is inaccurate and should be corrected.	The schedule of activities is vague, not logical, or omitted. Detailed instructions are not included for instructors or students. There are so many inaccuracies in factual information that this unit should not be taught.

Core components of a teachable unit:

- **Learning goals:** What will students know, understand, and be able to do?
- **Intended learning outcomes**: What performances or behaviors will indicate whether students have met the learning goals?
- **Assessment activities**: What activities will give students and instructors regular feedback about learning?
- **Classroom activities**: What activities will engage a diversity of students in learning?

Themes of a teachable unit:

- **Active learning**
- **Assessment**
- **Diversity**
- **Scientific teaching**

Scientific teaching and teachable units

Development of a teachable unit is not a one-time task; it is an ongoing process of review and revision that requires feedback. The most obvious—and arguably most important—feedback comes from the evaluation of the learning outcomes. Did students achieve what you had intended?

Scientific teaching adds another dimension to review, which is parallel to peer review that typifies scientific research. Feedback can originate with the instructor, colleagues, or students. Table 5.2 provides descriptions of criteria for the development of a teachable unit that can be used for review. The rubric is written with categories that describe "levels of completion" to emphasize that development of a teachable unit is an iterative process of review and revision.

Elements to consider when reviewing a teachable unit:

- **Scientific content**: How do the goals represent the concepts and nature of science?
- **Active learning**: How will the activities engage students in learning?
- **Assessment**: How will achievement of the goals be measured and provide feedback?
- **Diversity**: How will the activities engage a diversity of students in learning?
- **Alignment, review, and revision**: Are the activities and assessments designed to help diverse students achieve the goals? Based on evaluation and review, what changes will improve instruction and learning?

Building a Teachable Unit: Backward Design

Backward design provides the framework for developing a teachable unit (Wiggins and McTighe 1998). We have modified Wiggins and McTighe's original three-step model to include a fourth step.

1. **Identify desired results (learning goals).**
   What will students know, understand, and be able to do?
2. **Determine evidence for learning (learning outcomes and assessment).**
   What behaviors and performances will typify achievement of the goals? How will students and instructors gauge progress toward the learning goals?
3. **Plan learning experiences and instruction (activities).**
   What activities will engage a diversity of students in learning?
4. **Align goals, activities, and assessments.**
   Do activities and assessments help students achieve the learning goals?

Backward design mirrors the process that scientists use for research, but few conceive of teaching this way. Its salient feature is the focus on learning goals and outcomes, and using assessment and activities as a vehicle to achieve them. In addition, the iterative nature of review and revision is an important component of scientific teaching, so a fourth step—"**alignment**"—has been added to the process. The alignment step can also be viewed as review and revision. Table 5.3 organizes the components in a teachable unit into a simple framework based on backward design principles.

Table 5.3. A framework for constructing a teachable unit.

Learning Goals	Assessment	Activities	Alignment
What should students know, understand, and be able to do at the end of the unit? Do the learning goals represent the nature of science?	How will I determine whether students have met the learning goals? How will students assess their own learning?	What activities will engage a diverse group of students in learning?	Do the activities and assessments help students achieve the learning goals?

Step 1: Set Learning Goals

Set learning goals that emphasize learning

Learning goals should focus on what students need to *learn*, rather than what needs to be *taught*. Many of us are in the habit of teaching what *we* want to talk or think about, instead of thinking of the minutes we spend in class as belonging to our students. The class session is the time to stimulate interest and initiate learning. It is important to consider what strategies will capture the students' interest in the subject, motivate them to learn on their own, and help them construct new knowledge.

Set learning goals that represent the nature of science

The first step in developing a teaching plan is to establish learning goals that reflect the many facets of science. But there is so much to teach in science; no one can possibly teach it all. How does an instructor choose what to include? Ask:

- **What do I want my students to know, understand, and be able to do?**
- **What information is essential?**
- **What knowledge or skills are relevant to the subject area?**

The answers to these questions undoubtedly have many layers because specific goals are unique to each instructor, course, and group of students. For example, the detailed knowledge about specific molecules and chemical reactions that is needed to understand the Krebs Cycle during a biochemistry unit may be different from what students need to know one month later during a unit on bacterial physiology. In one course, it may be essential for students to build skills working in groups, while group skills may be ancillary in the next course. Not every course needs to address all the goals of the college biology curriculum, and not every course will award each goal the same attention. What matters is that the learning goals make explicit what understanding is critical during a particular course or unit at a particular moment in time, that the goals represent the nature of science, and that the goals link to the broader aims of the entire undergraduate curriculum. See Table 5.4 for an example of the breadth of learning goals.

Table 5.4. Examples of learning goals that represent the nature of science (National Research Council 1996)

Aspect of Science	Examples of Goals
Knowledge	Students will know key concepts and seminal facts in the discipline. Students will understand which techniques are used to answer which scientific questions.
Skills	Students will be able to design and conduct scientific investigations. Students will be able to think critically about how experimental evidence answers a scientific question. Students will be able to perform relevant laboratory techniques.
Behaviors	Students will be able to collaborate with other people. Students will be able to communicate and defend a scientific argument. Students will be able to implement a technological solution to solve a scientific or societal problem.
Attitudes	Students will be curious about the world and motivated to take responsibility for learning about it. Students will appreciate the relevance of scientific discovery in today's society and throughout history. Students will empathize with the hard work it takes to do scientific research.

Elucidate the hierarchy of learning goals

A clear structure and conceptual framework provides a way to organize information and prioritize the most important goals. The hierarchy of goals becomes the scaffolding on which the rest of the unit is built. Therefore, it is important to delineate what epitomizes lasting understanding and what is considered transient knowledge and to differentiate which goals are essential, important, interesting, or irrelevant.

Wiggins and McTighe (1998) describe a filter for clarifying instructional priorities. The first step is to determine which goals are **essential** to understand (for example, they have lasting value beyond the classroom), which are **important**

Primary conceptual goals
These are *essential* to understand.

↓

Secondary conceptual goals
These are *important* to understand.

↓

Specific topics
These *illustrate* what is to be understood or present *interesting* examples with which students should be familiar.

↓

Superfluous materials
These material should not be included in the lesson.

Figure 5.1 Filter for learning goals. When developing instructional materials, first categorize which goals are essential, important, or interesting. Jettison the superfluous materials.

(for example, they represent a core idea or process in a discipline), and which are **illustrative** or **interesting** (for example, they are worth being familiar with or are currently relevant).

Another way to look at learning goals is to consider the difference between topics and concepts. *Topics* are used to illustrate concepts and are more specific. Therefore, topics, carry different weight altogether from *concepts*. In addition, certain concepts are more integral to the subject area than others. Table 5.5 provides some examples of primary and secondary concepts and topics in biology.

Know that content matters

An important goal of any course is to ensure that students learn the scientific knowledge that is appropriate and expected in that discipline. Therefore, content must be accurate. This may sound obvious, but it is striking how often introductory texts and instructors provide inaccurate or outdated information, doing a disservice to the students, the instructor, the course, and to science. Science requires knowledge of memorized facts, figures, and principles.

Table 5.5a. Example of learning goals for a scientific concept (evolution)

Goal	Example
Primary concept	Students will understand that populations of organisms evolve because of variation and natural selection.
Secondary concept	Students will know that a mutation in DNA can lead to a change in protein structure and function.
Specific topic	Students will know that a mutation in the gyrA or parC genes of *Neisseria gonorrhoeae* bacteria can lead to antibiotic resistance and, therefore, increased fitness in the presence of the antibiotic ciprofloxacin.

Table 5.5b. Example of learning goals for a scientific process (experimentation)

Goal	Example
Primary concept	Students will understand the importance of data in scientific investigations.
Secondary concept	Students will be able to analyze experimental data and explain clearly whether the results support the hypothesis.
Specific topic	Students will be able to work together to design and conduct an experiment to determine the cause of leaf death in an apple orchard.

But facts are learned and remembered better when they are presented in a scientific context and when students have the opportunity to apply them. Thus, using concept-based learning that emphasizes application of knowledge does not mean abandoning the goal of learning facts, it means learning them *better*.

One resource that organizes biological concepts and topics into a hierarchy of learning goals is the online article by Khodor *et al.* (2004). The article can be used as a starting point for framing biological content goals for a course or as a model for establishing hierarchy among other learning goals. Another resource is *Bio2010* (National Research Council 2003), which provides recommendations for central themes in biology.

Consider what students already know and target the difficult concepts

Students bring knowledge to the classroom, and this affects what and how they learn. Instruction should be designed to figure out what students already know and build on it. Ask:

- **What prior knowledge, previously held ideas, or misconceptions might students have?**
- **What are the conceptually challenging aspects of this material?**
- **What experiences will encourage students to share their misconceptions so they can be addressed?**

Identifying these challenges will focus everything else in the unit. Content and teaching methods can be tailored to the challenge so classroom time and material can be invested where it will have the greatest outcomes. Educational journals are a great place to start; most common misconceptions in biology have been identified, and many correspond with teaching solutions (Pfundt and Duit 1994). By identifying these challenges, the instructor can choose to accept them or resolve them. For example, in a unit about plant biology, students probably know before the unit starts that plant cells have the capacity to carry out photosynthesis, but few realize that all plant cells respire.

Step 2: Determine Evidence for Learning (Learning Outcomes)

Determine what qualifies as acceptable evidence that students are learning

This step translates learning goals into specific outcomes that students should achieve, including their behaviors and performances. Students who understand a concept should be able to demonstrate their knowledge in multiple ways, and there are different models for articulating outcomes. By using the language and ideas from these models, instructors can translate learning goals into intended learning outcomes. For example, Wiggins and McTighe (1998) propose that there are six facets of understanding that should be used to describe learning outcomes: explanation, interpretation, application, perspective, empathy, and self-knowledge. Similarly, Bloom *et al.* (1994) provides a hierarchy of intellectual behaviors that demonstrate learning: evaluation, synthesis, analysis, application, understanding, and knowledge. Chapter 1 provides an overview and context for these and other learning models.

In establishing learning outcomes, it is important to consider:

- **What performance or behaviors indicate that students understand?**
- **What criteria will differentiate among the different levels of understanding?**
- **How will students know whether they are learning?**
- **How will I know whether students are achieving the learning goals?**
- **What assessment tools will engage students and measure learning?**

Once the outcomes are defined, assessment tools can be designed to measure student learning during a unit and at key endpoints. For example, "enGaugements" are activities that provide regular feedback about progress during the unit. Short-answer essay questions can assess multiple goals simultaneously at the end of a unit. Rubrics differentiate among the varying levels of performance that students achieve. Chapter 3 provides more comprehensive information about these types of assessment tools and how to use them effectively in instructional design. Keep in mind that feedback is one of the most important factors in increasing student learning gains (Wiliam and Black 1998, Hattie 1999), so assessment activities should be selected carefully. Table 5.6 provides a comparison of learning goals and outcomes.

Step 3: Plan Learning Experiences and Instruction

Choose activities to help students achieve the learning goals

Now is the time to gather all those creative activities and ideas together and determine which ones will help students achieve the learning goals. In addition, consider how activities can be used elicit the students' prior knowledge, build new knowledge, and address common misconceptions. The filter (from Step 1) provides a tool to evaluate which concepts should be emphasized with multiple or in-depth activities, and which concepts or topics can be covered in less depth. Key questions to ask are:

- **What activities will help students achieve the learning goals?**
- **Do the active learning exercises focus on the most important (primary) or difficult concepts?**
- **What material can students learn best in a lecture, discussion, or lab, and what could be learned just as well on their own or in groups outside of class?**

Consider the order of activities in the classroom

Learning goals and outcomes are important, as are the order of activities. Each event in the classroom should build on previous activities and knowledge. The

Table 5.6 Comparison of learning goals and outcomes.

Type of Goal	Learning Goal	Learning Outcome
Knowledge	Students will know the structures of the amino acids.	Students will be able to draw each amino acid and arrange them based on chemical properties.
Comprehension	Students will understand the role of genetic information in classifying organisms.	Students will be able to classify organisms based on 16s RNA sequence information.
Application	Students will understand the relationship between genes and proteins.	Students will be able to generate a mini-map using the following terms: DNA, RNA, protein, transcription, translation, amino acid.
Analysis	Students will be able to analyze data.	Students will be able to calculate the frequency of Proteobacteria in insect guts compared to other types of bacteria.
Synthesis	Students will understand how to design an experiment.	Students will be able to formulate a hypothesis about solute transport and design an experiment to test it.
Evaluation	Students will understand the importance of invasive species in ecosystems.	Student groups will develop a plan to deal with an invasive wetland species and defend it in an oral presentation to the entire class.

"5E" model of instruction provides a logical way to think about the order of activities within a teachable unit (Bybee 1993, Eisenkraft 2003). Below is a modified version.

- **Engage and Elicit:** Students are motivated to learn by connecting a new challenge with prior knowledge and experiences.
- **Explore:** Students actively experience the knowledge and tools needed for the topic, connecting concepts and skills.
- **Explain:** Students articulate what they discovered during the exploration.
- **Elaborate:** Students practice skills and vocabulary while deepening their conceptual understanding; this step may require many iterations.
- **Evaluate:** Students and instructors determine whether learning goals are being met throughout the unit and at the end of the unit.
- **Extend:** Students apply their knowledge or skills to a new situation.

And consider what activities can occur outside the classroom

Much of learning occurs *outside* the classroom, so classroom time should be carefully constructed to motivate students, frame the concepts, and identify challenges. We might forgo showing the students the structures of all the amino acids, and simply ask them to memorize the structures outside of class. Instead, we could use class time to discuss the chemical properties of amino acids and then ask the students to use the principles in solving problems about protein chemistry to help them apply the structural information and understand why it is important.

Build on instructional resources that already exist

There is no need to reinvent the wheel! Instructional materials abound on the Internet and in print. For activity ideas, use resources from this book, delve into online "digital libraries," or talk with colleagues about activities they have tried. For example, digital libraries contain vast amounts of freely available, peer-reviewed instructional material that can be used as is or adapted to meet specific learning goals. Broad searches often turn up useful, innovative materials in other disciplines.

Vary the content and examples (for the purposes of inclusivity and diversity)

Part of the thrill of teaching is to share what we know best, so we choose examples based on what we know. One liability of this strategy is that the subject matter itself can be inclusive to some students and exclusive to others. Sometimes the examples we choose can be exclusive to students in subtle ways never intended or considered. A single illustration may not be offensive to anyone, but over the course of a semester in many classrooms, repeated use of certain metaphors or types of people can be alienating to some students. For example, female and minority students who never see people who look like them represented in images or examples during the semester may begin to feel disenfranchised. Sports, military, or mechanical metaphors or analogies that may only make sense (or be interesting) to certain members of the class can have the same effect. Changing the examples doesn't take any more time in the classroom, and the benefit is to make more diverse students perceive themselves as part of the scientific community.

Step 4: Alignment

Check alignment

The first three steps in this chapter describe how to use backward design to approach the development of a teachable unit (Table 5.3). The final step is "alignment," which ensures that the activities and assessments are designed to help students meet the learning goals.

During this step, it is important to ask:

- **Do the activities help a diversity of students achieve the learning goals?**
- **Do the assessments drive learning toward the goals and provide ample opportunity for students and instructors to gauge progress toward—and achievement of—the learning goals?**
- **What revisions would help the activities and assessments align better with the learning goals?**
- **Do the learning goals represent the nature of science?**

Review and revise

Alignment highlights the concept that instructional material design is an iterative process. Even though the process is described as linear, it really is a cyclical process that requires reflection and revision at every stage. If the activities and assessments don't help students meet the learning goals, then they should be revised accordingly. Often, the learning goals become more refined during this process, too. As you review and revise goals, jettison superfluous materials and experiences that do not help students meet the learning goals. In short, ensure that learning experiences align with the learning goals.

CHAPTER 6

Institutional Transformation

Institutional Transformation Defined

Institutional transformation is the process of transforming the culture and practices of a campus to reflect a commitment to key ideals. In scientific teaching, this means cultivating enthusiasm for the intellectual endeavor of teaching and developing practices that recognize and reward teaching innovations and successes.

Why Institutional Transformation?

Faculty who adopt scientific teaching may be content to know that their teaching is based on sound principles, but some faculty intend to effect change beyond their own classrooms. Some want the respect of their colleagues instead of receiving blank stares or belittling comments when they discuss their new teaching approaches. Others need resources to buy equipment or modify their classrooms in order to implement their teaching plans, but their colleagues and administrators have no idea why these changes are needed and do not appreciate their value. Others would like scientific teaching methods to be used throughout a course series so that students become used to the teaching methods. Still other faculty want to share their excitement and successes with colleagues to improve teaching throughout their institution. But left to their own devices, institutions of higher learning change very slowly.

Models for Change

Stages of change

Institutional change has been compared with human behavioral change (Prochaska and DiClemente 1983, Prochaska *et al.* 2001). Physicians have delineated the "five stages of change" in human behaviors that patients move through on their way to change health-related behaviors involved in smoking, eating, or exercising. The five stages are described in Figure 6.1.

Precontemplation → Contemplation → Planning → Action → Maintenance

Figure 6.1 The Stages of Change Model (Carnes *et al.* 2005)

Every campus and the individuals on it are at different stages in this process of change. Some may be in the **precontemplation** stage, still believing that the way we teach is just fine, ("After all, I learned a lot, didn't I?"). These colleagues will need to be convinced with different tactics and information than those colleagues in **contemplation** ("I guess I don't know whether my teaching methods are effective") or **planning** ("I want to make changes, but I'm not sure how to implement active learning") stages.

We present the stages of change in a linear model, but just like everything else in scientific teaching, it really is an iterative process of evaluation and revision, of progress and stasis. Feedback, collaboration, and metacognition are hallmarks of each of these models.

Common myths

As you have probably realized from reading this book, there is no single best way to approach teaching, but certain principles and practices underlie every chapter. Recurring themes in scientific teaching include: integrating learning theory into practice; incorporating the rigor and spirit of scientific scholarship; and using methods that have been tried, tested, and shown to be effective. It is important to be familiar with these themes to be effective at implementing changes; it is equally important to know some common myths so they can be replaced with accurate information.

Table 6.1 describes common myths about scientific teaching concepts, including the definitions of each term based on their use in this book. Some of the myths are based on inaccurate information ("Students' minds are empty vessels"), and some myths are subtle, complex, or only partially wrong ("Students should be self-motivated to learn"). Some of the ideas are alluring in their simplicity ("If I'm not lecturing, then I'm not teaching" or "I treat all students the same in my class") and others feel uncomfortably true for almost everyone ("There are too few people with too little time to make real change at this campus"). It's up to the reader to recognize when these myths arise, and to decide whether to implement strategies for change.

Institutional Transformation in Practice

Instructors and administrators often want guidance to accelerate and measure the rate of institutional change. This chapter presents strategies for instructors and administrators to effect changes—small and large—at the local and global levels. A few themes pervade these recommendations. First, data are essential to institutional transformation. Evidence-based arguments are received better by scientists and are easier to defend than opinions, so it is critical to use both review articles and original research studies that document the value of scientific teaching (Handelsman *et al.* 2004). Data collected from departments on the campus undergoing transformation are especially powerful: it's hard to argue against the fact that students right there on one's own campus (not some theoretical students in the literature) are learning less than they might be if they were taught differently. Finally, enthusiasm and positive examples propel any movement forward. It is hard to dismiss a colleague who is armed with good values about learning goals, data about student outcomes, excitement, and examples of success.

Campus-wide policies

A pervasive myth at research universities is that teaching doesn't matter, thereby steeping faculty in a culture that places teaching in conflict with research. Consequently, policies, when reinforced by dedicated administrators, can transform

Table 6.1 Definitions and Common Myths about Scientific Teaching

Topic	Definition	Common Myths
Scientific Teaching	Teaching science in a way that represents the true nature of science and approaches teaching with the rigor of scientific research.	Scientific teaching requires extensive understanding of educational literature and assessment techniques. Undergraduate students are not sufficiently sophisticated to understand scientific inquiry; inquiry is learned in graduate school. If I'm not lecturing, then I'm not teaching.
Active Learning	The process in which students are actively engaged in learning.	Active learning takes too much time and occurs at the expense of learning content. I have to cover the content for the next course. Content must be covered at all costs. Other goals are secondary or irrelevant. As long as students are busy, they're learning. Students' minds are empty vessels or blank slates. If students are not taking notes, then they are not learning.
Assessment	Measuring progress toward and achievement of the learning goals.	The point of assessment is not to help students learn—the point is to measure what they have learned. I know I'm successful; the students who return say they remember my teaching and how it affected them. I don't have the background in assessment needed to be a scientific teacher.
Diversity	The characteristics that make each student unique, each cohort of students unique, and each teaching experience unique. Diversity includes everything in the classroom: the students, the instructors, the content, the teaching methods, and the context.	Students should learn to study the way I did. Students should be self-motivated to learn. It's the student's job to achieve, not mine. Culture, ethnicity, gender, and background have no place in the classroom. It's simply about learning the facts.
Institutional Transformation	The process of changing the culture of a campus to reflect a commitment to—and practices aimed at—improving undergraduate science education.	There are too few people with too little time to make real change at this campus. Student evaluations are the only measure of teaching success. Teaching accomplishments don't matter on my campus.

campus values and practice. Policies can elevate the importance of teaching as a whole and make clear that the administration values the integration of teaching and research—rather than viewing teaching and research as separate, competing endeavors. For example, allocating 1/3 of merit pay increases to instructional practices sends a message that teaching matters and that it will be rewarded. Tenure guidelines can include strong language to indicate that teaching accomplishments are necessary for tenure and should include a clear description of what teaching criteria will be evaluated. Sabbaticals can be awarded based on proposals to improve teaching or to integrate teaching and research, rather than on research-only plans. Some administrators offer additional new-faculty start-up funds that are designated for teaching. These funds can be used to attend teaching workshops, purchase equipment for student-friendly classrooms, or host educators from other universities to present seminars on teaching.

Publicity and marketing

Administrators can use their bully pulpits to promote the integration of teaching and research and the value of teaching. Presidents and provosts can discuss teaching in speeches, highlighting innovations and why they matter. They can instruct their campus newspaper staff to cover education topics and feature faculty who are recognized for their research as well as teaching. Demonstrating that respected researchers are dedicated to and excel in teaching can influence other senior faculty and sends the message to junior faculty that teaching is held in high esteem by the university's most influential researchers.

Deans can advertise their decisions to allocate resources based, in part, on teaching effort and quality, providing funds to support innovation and revamping of old curricula. Department chairs can use these policies to induce their faculty to engage in teaching reform. It is essential that chairs make clear to the entire faculty that the efforts of innovators reflect well on the entire department; when the other faculty notice that their department is recognized and rewarded for teaching innovation, they are more likely to become involved in the reform movement. At the least, they will appreciate the efforts of those who are engaged in the reform.

Professor-to-professor dissemination

Most instructors who engage in education reform are eager to share their new-found knowledge with colleagues. Many are dismayed to find that their colleagues look askance at the new teaching methods and are dismissive in conversations about teaching. Changing faculty behavior is never easy. It can feel Sysiphean—endlessly repetitive, unrewarded, and defeating. But a few strategies have been found more effective than simply telling colleagues about new techniques, which is bound to fail. One of the most effective is team teaching. When colleagues see active learning in action, they can't deny its power. Students will aid the case because once they become used to active engagement, they will complain about instructors who force them back into the passive mode. Another way to engage colleagues, letting the methods speak for themselves, is to invite peer review. Most colleagues are flattered by being asked to review another's teaching and everyone benefits: the instructor being reviewed will often gain useful insights, the reviewer will see active learning in action, and the two instructors will begin a dialogue about teaching. Finally, departmental chairs can invite their faculty to read a book or article about teaching, such as *Leaving the Lectern* (McManus 2005) and discuss it at a faculty meeting or faculty retreat.

Campus teaching events

Well-constructed teaching seminars and courses can be transformative. Colleagues who attend them are no doubt far along the change continuum—already in the contemplation stage or beyond and convinced that some change is needed—so this is not a strategy that will work for all colleagues. But for those who are interested and want to learn, this can be a high impact way to reach many instructors. In addition to seminars for faculty, courses for graduate students and postdocs who will be the next generation of faculty are essential ingredients in the recipe for change. In addition to developing good teaching practices themselves, they can be powerful ambassadors for change by informing their advisers and colleagues about what they are learning, thereby enlarging the pool of people affected by the course beyond just those attending it. The following section describes workshops that are designed to train faculty and future faculty in scientific teaching and to empower them to be agents of change.

SCIENTIFIC TEACHING WORKSHOPS

Scientific Teaching Workshops Defined

The Scientific Teaching Workshops are intended to assist instructors in effecting broad-scale change by introducing others to the principles and practices of scientific teaching. The workshops are designed to

- help participants develop teaching skills or experience new teaching methods
- model the principles and techniques that are being discussed
- target concepts that may be difficult to understand
- foster discussion among participants
- build on the collective experience and knowledge of the participants
- cultivate a community of sharing and peer review in teaching
- provide a foundation for institutional transformation in science education

Why Workshops?

The Scientific Teaching Workshops provide strategies to explore scientific teaching in depth through a diversity of activities and assessments and to induce change in science education. The goal of the workshops is to provide participants with experiences and discussions that solve knotty problems in teaching and build community and respect for teaching innovations. Each workshop models the principles and practices for each component of scientific teaching.

Workshops in Practice

Workshop Structure

Est. Time	Topic
5 min	Overview and goals
35–95 min	Cases, activities, and discussion (Actual time depends on the number of activities)
15 min	Action plans and summary
5 min	Exit assessment

Format of the workshops

The workshops can stand alone as independent events such as a one-time seminar, or they can be integrated into a series of ongoing events. For example, the workshops could serve as the topics to foster departmental discussions about improving teaching in a "brown-bag" lunch series. Alternatively, the workshops can be linked together formally in a graduate-level course that centers on building a teachable unit and includes more in-depth activities like Reading Assessments (see chapter 3) to help train a new generation of faculty. As a series, the workshops can help participants build a wealth of knowledge, strategies, and innovations in teaching that can be used to work toward institutional transformation in science education. Because they are designed to build on the collective experience and knowledge of the participants, the cohort of participants learns to address problems collectively and intellectually, rather than personally and individually. Consequently, it becomes the responsibility of the entire group to improve teaching.

Facilitation tips

The collective wisdom of the participants is the heart of the content for the workshop. Whenever possible, participants should ask their own questions and answer each others' questions. The facilitator's most important role is to foster discussion: to get the discussion going, keep it on track, encourage reluctant participants, and gently quell zealots. Therefore, the facilitator should not dominate the session with a lengthy lecture or personal experiences. Each workshop description includes guiding questions for each workshop and suggestions for how to model 1-2 concrete examples of each topic. Examples can come from your own experience, from a colleague's experience, or from examples in each chapter of this book.

Preparation

To prepare for a given workshop, facilitators should read the appropriate chapter in this book. Each chapter provides background information on core concepts, and it includes tables and figures that can be used to illustrate the ideas during the workshop. In addition, consider which guiding questions will help you to address the key aspects of each workshop and whether the participants would benefit from having a copy of the appropriate "Resources" page.

Getting participants

Participants can be faculty, instructors, graduate teaching assistants, postdocs, or administrators. Each workshop can be easily adjusted for a particular audience. Whenever possible, try to construct heterogeneous groups which tend to have more complex and dynamic discussions. For example, consider the diversity of ages, career stages, ethnicity, or gender. You may want to offer the workshop first in your own department or program and then offer it more broadly through existing campus teaching and learning groups.

Evaluation

Embedded in each workshop is an exit assessment activity to provide the facilitator with some data on participant learning. Facilitators who are interested in collecting additional data such as demographics of the workshop participants, satisfaction, and overall effectiveness are encouraged to survey participants more extensively.

Institutional support and future directions

Teaching evaluations of the workshops tend to be very positive. Facilitators may want to lead one or more workshops and use the evaluations to enhance a tenure/promotion package, to recruit departmental colleagues, to mentor a junior faculty member, or to build a case for campus-wide adoption of the seminar.

Overview of Scientific Teaching Workshops

Workshop Title	Goals
Scientific Teaching	Understand that scientific teaching refers to teaching science in a way that represents the true nature of science and to designing teaching approaches based on evidence. Be able to identify strategies that reflect scientific teaching.
Active Learning	Understand that active learning engages students in thinking and behaving like scientists. Be able to choose teaching methods that actively engage students in learning.
Assessment	Understand that assessment drives student learning, it is more than grades, and it provides feedback to both instructors and students about learning. Be able to develop assessment tools that provide frequent feedback to both instructors and students and that evaluate progress toward and achievement of the learning goals.
Diversity	Understand that diversity affects learning; that knowledge, experience, and cultural influences make each student unique; that the teaching methods, examples, and content we choose to use affect who is included or excluded from the classroom experience; and that we all bring biases and assumptions to teaching and learning. Be able to choose teaching methods that create inclusive classrooms.
Institutional Transformation	Understand the importance of building a community of support in teaching and learning, leveraging administrative support and funding, and giving/receiving recognition for teaching accomplishments. Be able to effect change on campus.

	Suggested Activities	Exit Assessment
	1. Discuss case: "Frustrated Professor" 2. Introduce scientific teaching: Mini-lecture 3. Brainstorm: Reasons for teaching science 4. Think-pair-share: Evidence for teaching	What is scientific teaching, and when have you seen it in practice?
	1. Discuss case: "Constructing Knowledge" 2. Introduce active learning: Mini-lecture 3. Think-pair-share: Learning gains in active versus passive lectures 4. Model examples of active learning techniques 5. Discuss: Process of active learning 6. Complete worksheet: "Active Learning" 7. Design an active learning exercise	Explain how active learning models the ways that scientists think and behave.
	1. Discuss case: "Frustrated Student" 2. Introduce assessment: Mini-lecture 3. Give examples: EnGaugements 4. Complete worksheet: "Compare and Contrast Two Assessments" 5. Complete worksheet: "EnGauge Students in Learning" 6. Discuss case: "Grading"	Explain how an assessment technique can simultaneously engage students in learning and help them gauge what they understand.
	1. Discuss cases: "Cousins Vang," "Inclusive Classrooms," and "Bantering" 2. Introduce diversity: Mini-lecture 3. Think-pair-share: Attrition statistics 4. Model examples of diversity in the classroom 5. Complete worksheet: "Creating an Inclusive Classroom"	How does diversity affect learning?
	1. Discuss cases: "Teaching Evaluations" and "Overextended" 2. Introduce institutional transformation: Mini-lecture 3. Think-pair-share: Common misconceptions about teaching 4. Share examples and ideas for institutional transformation 5. Brainstorm and discuss strategies	What would a "transformed" department look like? What is your role in changing science education at this campus?

Workshop Contents

Leading the Workshops as a Series or Graduate Course

Overview and Goals

Individual workshops can have a significant impact, and they serve as a low-threshold way to introduce the individual ideas of scientific teaching, active learning, assessment, diversity, and institutional transformation to a broad audience. However, offering these workshops in a series can be much more influential because they have a concerted effect. Of particularly high impact is a series that engages colleagues in the process of developing instructional materials by integrating the approaches from all the scientific teaching workshops.

This workshop series can be taught in several formats. For example:

Scientific Teaching Institute. The workshops can be offered as an intensive one-week institute about scientific teaching. The workshops provide the formal sessions of the institute, and the remaining time periods could be used to work on building, reviewing, and presenting a teachable unit.

Weekly Scientific Teaching Seminar Series. The workshops can provide a seminar series within a department. For example, Tuesdays at noon, faculty and instructional staff can convene for a workshop, alternating with weeks to discuss difficult teaching issues or develop new instructional materials with the group's feedback.

Graduate Course. The Scientific Teaching Workshops can be the basis for a graduate course. The workshops provide a foundation that graduate students can use as the basis to develop a teachable unit that is peer reviewed. In the following semester, the graduate students can implement their materials in an undergraduate classroom, evaluate them, revise them, and disseminate them.

Whatever the format, a central activity of the workshop series is development of teaching materials based on scientific teaching principles, which engages participants in scientific teaching. As the participants work together to learn about and develop instructional materials, they create community and synergy in teaching.

The specific elements of a teachable unit and the evaluation criteria for each of these components can be found in Chapter 5 of this book. Implementation, evaluation, revision, and dissemination of the teachable units generated from workshop series can vary based on the needs of your institution.

A Typical Session

Each workshop is designed to last approximately two hours. The time can be varied based on the number of activities included in each session. Following an introductory workshop on Scientific Teaching, the progression of the workshops follows the steps of developing the teachable unit using backward design: setting learning goals, assessment, active learning, and alignment. The specific activities for each of these workshops are shown in the table about "Building a Teachable Unit: Workshop Series" and in most cases are described in detail in each of the workshop sections. As a course, the first hour can be filled with a "Reading Assessment" activity (see description, below), followed by an hour of the workshop materials.

Reading Assessment: Participants are asked to read the selected reading before the session and 2–3 participants are asked to prepare a reading assessment for that session. The primary goals of the reading assessment are (1) to engage participants in an intellectually challenging discussion about a reading assignment (2) to assess whether the participants understand the reading or are able to apply it to their own teaching, and (3) to share the teaching responsibilities for each session and give participants time to practice teaching with new methods. Importantly, reading assessments are designed and led by the workshop participants to provide them with practice teaching their peers. In each session, participants should link the major concepts of the reading to the development of their teachable unit. Examples of reading assessments are included in chapter 3 with more information and guidelines.

Activities and Discussions: Select 1–3 activities from each workshop section. Alternatively, develop an activity of your own to engage participants in learning the major principles of scientific teaching. In each session, structure the activities so that participants can apply what they have learned directly to their teachable unit. Make sure there is ample time for participants to share progress on their teachable units and receive constructive feedback.

Assignments: Because the workshop series is focused on the development of a teachable unit, it is important that participants work on certain elements of the teachable unit between each session. Some appropriate assignments are suggested in the table "Building a Teachable Unit: Workshop Series" near the end of this chapter.

Tips for Facilitators

It is important for the facilitator(s) of this workshop series to guide learning without having the participants rely solely on the facilitator for knowledge. The most successful sessions will be those in which participants rely on each others' collective knowledge and experiences.

When teaching the workshop in a series, some tips are:

1. Keep in mind the goals for the entire series and not just the goals for each workshop.
2. Identify the key points you want participants to learn from each session, then try to find ways to help participants discover them on their own.
3. Reinforce the importance of collective problem solving in finding multiple solutions for a single issue.
4. Help participants define the scope of their teachable units early in the series.
5. Use the participants' collective experiences to predict what challenges might arise in the classroom when implementing each teachable unit.
6. Consider whether participants will practice teaching parts of the teachable unit to each other.

Evaluation

In-class activities and assessments: Each activity is designed to provide feedback about learning to the participants and the facilitator.

Exit cards: These simple, one-minute papers provide a snapshot of what the participants have learned and what they are struggling to understand.

Peer review committee: An important part of scientific teaching is review by colleagues. Participants in the workshop series may become each others' peer-review committees, or they may choose to include other colleagues (experts in teaching or the scientific subject area, or even undergraduates) to provide review. Peer-review committees can provide important evaluation during the preparation and presentation of the teachable unit.

Rubric: The "Teachable Unit Review Rubric" in chapter 5 provides guidelines for participants to self-evaluate their progress toward a comprehensive teachable unit, and for others to provide constructive feedback as well.

Mid-course check-in and exit survey: It is important to check in with participants mid-way through the series and at the end of the series using informal questions and formal surveys. One example of a formal survey is the Student Assessment of Learning Gains (1999b). Consider asking questions to address multiple aspects of the series.

Sample Mid-course Survey

1. Please comment on the effectiveness of the following aspects of the course (or workshops) so far:
   - Content of the sessions
   - Format of the sessions
   - Time involved
   - Facilitator(s)
2. How is your teachable unit progressing? What do you feel you have accomplished so far? What issues or questions do you have about your unit?
3. Do you feel that you are making progress toward the learning goals of the course? Please explain.
4. What are the three most important things you have learned so far?
5. What suggestions do you have for mid-course changes?

Building a Teachable Unit (TU): Workshop Series

Topic	In-class Activity	Assignment *Due in class*	Reading *Due in class*	Goals *Participants will. . .*
Scientific Teaching	Share goals and highlight which to achieve during this series. Brainstorm reasons we teach science.	Write a teaching philosophy	*Scientific Teaching* (Chapter 1) *Scientific Teaching* (Chapter 5: Read to Step 1) Instructional materials databases: www.merlot.org www.nsdl.gov	Understand main concepts in backward design and scientific teaching Establish teaching goals for the semester and brainstorm strategies to achieve them and measure success
Learning Goals	Share goals for peer review Apply National Science Education Standards and Biology Concept Framework to the development of a Teachable Unit (TU)	Write learning goals for the TU Describe what prior knowledge is expected Describe important misconceptions in this subject area and explain how the TU addresses them	*Scientific Teaching* (Chapter 5: Step 1) National Research Council 1996 (Chapter 6) Khodor *et al*. 2004	Apply backward design: Establish overall goals and content goals Evaluate whether learning goals represent nature of science Review and revise goals
Learning Outcomes and Assessment	Share assessments for peer review Define outcomes for TU List the tools you will use to measure outcomes Explain how the assessments engage students and help them gauge their learning	Describe outcomes and list assessment tools for the TU Build a rubric for the TU	*Scientific Teaching* (Chapter 5: Step 2) *Scientific Teaching* (Chapter 3) National Institute for Science Education 1999a	Apply backward design: Describe what outcomes are expected and design tools to measure outcomes Review and revise assessment tools
Active Learning	Share activities for peer review Apply modified version of the "5E" model to your activities Explain roles of student and	Describe activities that students will do Mid-course check-in	*Scientific Teaching* (Chapter 5: Step 3) *Scientific Teaching* (Chapter 2)	Apply backward design: Describe what activities students will do and create a schedule for the classroom

Alignment	Share goals for peer review Describe how activities and assessments help students meet the learning goals	Explain how the assessments and activities help meet the learning goals Explain how you are meeting your goals so far	*Scientific Teaching* (Chapter 5: Step 4)	Determine whether activities and assessments align with goals Evaluate whether you are progressing toward your teaching goals and describe strategies for next phase
Diversity	Share goals for peer review Explain how the activities and assessments enGauge all students in learning Apply diversity questions from workshop #4	Explain what strategies you have used to reach all students Explain how you will determine whether students are being included or excluded	*Scientific Teaching* (Chapter 4)	Determine whether all students will be included in learning Review and revise diversity approach
Presentations	Present part of the TU Give and receive feedback	Prepare to teach a representative portion of your TU Review a peer's TU before presentation and receive feedback on your TU	*Scientific Teaching* (Chapter 5: Teachable Unit Review Rubric)	Chance to practice presenting materials, leading activities, and gathering assessment data Learn to give and receive peer review in large groups
Scientific Teaching	Read each other's teaching philosophies and discuss Complete the TU	Fill in details of activities and assessments, including rubrics (comprehensive TU) Write instructions and tips for other instructors Include a schedule of activities (for both instructors and students)	Each others' teaching philosophies	Evaluate whether teaching goals have been met Understand that scientific teaching and instructional design is a process of review and revision based on goals and outcomes

WORKSHOP I

Scientific Teaching

Workshop I: Scientific Teaching

Topic	Suggested Activities	Goals Participants will...
Welcome & Overview	Introduce goals and schedule of the workshop.	Know schedule and context for the workshop.
Scientific Teaching	1. Discuss case: "Frustrated Professor" 2. Introduce scientific teaching: Mini-lecture 3. Brainstorm: Reasons for teaching science 4. Think-pair-share: Evidence for teaching	Engage in learning about and practicing scientific teaching. Share prior knowledge, experience, and misconceptions about scientific teaching. Analyze one professor's approach to teaching. Understand that scientific teaching is an ongoing process that makes teaching more scientific. Explore scientific teaching in depth and consider ways to apply it to teaching.
Summary	State key points of the workshop. Share resources. Complete the action plans.	Understand key points of the workshop. Have tools to apply scientific teaching in practice. Reflect on the workshop.
Exit Assessment	Have participants write for one minute: *What is scientific teaching? Give an example.*	Evaluate what was learned during this workshop.

Workshop Preparation

- Review Chapter 1.
- Review the cases, activities, and guiding questions.
- Make copies of the cases, resources, and action plan for participants.

Workshop I: Scientific Teaching

Notes and Tips for Workshop Facilitators

I. Goals

The goal of this workshop is to help participants understand that scientific teaching refers to **teaching science** in a way that represents the nature of science and to the **science of teaching**.

By the end of this workshop, participants will

- ▶ Understand that a scientific teacher sets learning goals that represent the nature of science.
- ▶ Have explored the breadth of reasons why we teach science.
- ▶ Be able to define what constitutes the nature of science.
- ▶ Understand that a scientific teacher evaluates learning regularly, makes teaching decisions based on evidence, and builds a community of teachers.
- ▶ Understand that scientific teaching is an iterative process of review and revision.
- ▶ Understand that active learning, assessment, and diversity are core themes of scientific teaching.
- ▶ Have developed a list of resources that can be used in the classroom.
- ▶ Have developed an action plan for scientific teaching.

II. Activities

1. **Discuss case: "Frustrated Professor"**

   Participants read and discuss the case.

2. **Introduce scientific teaching: Mini-lecture**

   Use chapter 1 for ideas and information. Give a few examples of scientific teaching in practice from chapter 1, your own experience, or an anonymous colleague's experience.

3. **Brainstorm: Reasons for teaching science**

   As a large group, participants brainstorm reasons to answer the question: *Why do we teach science?*

4. Think-pair-share

Participants work individually to answer the question:

What evidence would convince you that students are learning the true nature of science in a course?

Follow up with a discussion about the question:
What evidence would convince you that a colleague is a good teacher?

Exit Assessment

What is scientific teaching? Give an example.

Case: Frustrated Professor

Before the semester started, I worked really hard to set goals for the course. During the semester, I have been covering the content in clear, efficient lectures that I think are really well-organized, but the students don't seem to be learning the material. In fact, 40% of students failed the first exam.

Students these days don't know how to take notes and study. They just don't get it.

Key questions to guide discussion:

- What issues might be contributing to this situation?
- Has the professor done her job?
- Have the students done their jobs? What challenges might they be facing?
- What suggestions do you have for the professor?
- Have you faced a similar challenge?

Scientific Teaching: Questions to Guide Discussion

1. What are your goals when you teach? Do they represent the true nature of science? Do you have goals for students—what they should learn—as well as yourself? What other goals shape the classroom experience?
2. What role do students and teachers play in your classroom? Who takes the responsibility for learning?
3. In one or two sentences, describe your teaching philosophy. What evidence do you have that you teach in a way that is consistent with your philosophy? What further evidence would you like to gather?
4. Can assessment of student learning help guide what teaching methods you use?
5. What teaching methods are needed in scientific teaching?
6. What is the role of assessment in student learning? In your classroom, do the assessments test what you really want students to know, understand, and be able to do?
7. Explain the role of colleagues, support, and peer review in scientific teaching.
8. Scientific teaching is a process, not a product. What does this mean?

Scientific Teaching Workshop Resources

Key Concepts in Scientific Teaching

Scientific teaching refers to **teaching science** in a way that represents the true nature of science and to the **science of teaching**.

- A scientific teacher sets learning goals that represent the complex nature of science.
- A scientific teacher chooses activities that engage a diversity of students in thinking and behaving like scientists.
- A scientific teacher evaluates learning regularly and makes teaching decisions based on evidence.
- Scientific teaching involves building a community of teachers.
- Scientific teaching is an iterative process of review and revision.
- Active learning, assessment, and diversity are core themes of scientific teaching.

Suggested Reading

- Handelsman, J., S. Miller, C. Pfund. 2007. Chapter 1, Scientific Teaching in *Scientific Teaching*. New York: W.H. Freeman & Co.
- Handelsman, J., D. Ebert-May, R. Beichner, P. Bruns, A. Chang, R. DeHaan, J. Gentile, S. Miller Lauffer, J. Stewart, S. M. Tilghman, and W. B. Wood. 2004. Scientific teaching. *Science* 304:521-522.
- National Research Council. 2003. *Bio2010: Transforming undergraduate education for future research biologists.* Committee on Undergraduate Biology Education to Prepare Research Scientists for the 21st Century, Board on Life Sciences, Division on Earth and Life Studies. Washington, D.C.: National Academies Press.

Action Plan for Scientific Teaching

Which tools, strategies, or resources from today's workshop will be most helpful to you in teaching this semester or in the near future?

Think of your last few lectures. What topics have you struggled to teach?

List two colleagues who could help you brainstorm strategies to address the struggle.

1.

2.

Next time you teach, try one of the strategies that you and your colleague identify. How will you know if the strategy is successful? What will you observe or measure?

WORKSHOP II

Active Learning

Workshop II: Active Learning

Topic	Suggested Activities	Goals Participants will...
Welcome & Overview	Introduce goals and schedule of the workshop.	Know schedule and context for the workshop.
Active Learning	1. Discuss case: "Constructing Knowledge" 2. Introduce active learning: Mini-lecture 3. Think-pair-share: learning gains in active versus passive lectures 4. Examples of active learning techniques 5. Discussion: the process of active learning 6. Complete worksheet: "Active Learning" 7. Design an active learning exercise	Engage in learning about active learning. Share prior knowledge, experience, and misconceptions about active learning. Analyze knotty teaching problems. Understand that active learning engages students in thinking and behaving like scientists. Explore active learning in depth and apply it to teaching. Be able to convert passive lecture into active lecture.
Summary	State key points of the workshop. Share resources. Complete the action plans.	Understand key points of the workshop. Have tools to apply active learning in practice. Reflect on the workshop.
Exit Assessment	Have participants write for one minute: *How do active learning techniques engage students in thinking and behaving like scientists?*	Evaluate what was learned during this workshop.

Workshop Preparation

- Review Chapter 2.
- Review the cases, activities, and guiding questions.
- Make copies of the cases, resources, and action plan for participants.
- Make copies of the active learning worksheet.

Workshop II: Active Learning

Notes and Tips for Workshop Facilitators

I. Goals

The goal of this workshop is to help participants understand that active learning techniques engage students in thinking and behaving like scientists.

By the end of this workshop, participants will

- Understand that students learn when their current views (prior knowledge and misconceptions) are challenged by new information, and they have to construct new explanations to account for the information.
- Understand that active learning provides a model for how scientists think and behave.
- Understand that students who engage in challenging activities are more likely to take responsibility for learning.
- Understand that group process and cooperation are integral components of active learning.
- Be able to define what constitutes an "active learning exercise".
- Be able to solve knotty teaching problems using active learning techniques.
- Have developed a list of tools, strategies, and resources that can be used in the classroom.
- Have developed an action plan for active learning.

II. Activities

1. **Discuss case: "Constructing Knowledge"**
   Participants read and discuss the case.
2. **Introduce active learning: Mini-lecture**
   Use chapter 2 for ideas and information.
3. **Think-pair-share.**
   Participants survey the results from published papers comparing the learning gains in active vs. passive lectures (Figs 2.1 and 2.2 in this book). Individually, propose a hypothesis to explain the gains in learning. Discuss with a partner.

4. **Examples of active learning techniques**
   Model two or three active learning exercises from chapter 2. Workshop participants should play the role of the students.
5. **Discussion**
   After leading the active learning exercises, discuss the process.
   a. What did the participants experience or learn?
   b. How does it compare to a traditional lecture?
   c. How do active learning techniques engage students in thinking and behaving like scientists?
   d. What are some benefits and drawbacks of using active learning techniques?
6. **Active learning worksheet**
   Instructions for participants: Work with a partner to convert passive transmission of information through a lecture into an active learning exercise that accomplishes the same goals.
7. **Design an active learning exercise**
   Instructions for participants: Identify a common misconception about evolution (or another topic). Design an activity to target that misconception.

III. Exit Assessment

How do active learning techniques engage students in thinking and behaving like scientists?

Case: Constructing Knowledge

I really struggle to teach evolution. Students seem to get lost in the details and miss the really big concepts like preexisting variation in a population, natural selection, reproduction, and change in gene frequency in a population. How can I make them understand the importance of these concepts? No matter what I say, many of the students' answers are Lamarckian.

Key questions to guide discussion:

- What issues might be contributing to this situation?
- How could active learning techniques help?
- What roles do the students' prior knowledge and misconceptions play in their ability to learn in this case?
- What suggestions do you have for the professor?
- Have you faced a similar challenge?

Active Learning Worksheet

With a partner, convert each passive lecture into an active lecture.

Passive Lecture	Active Lecture
Every cell in an organism has the same DNA, but different genes are expressed at different times and under various conditions. This is called gene expression.	
Different parts of your body can do different things. For example, your hand has fine motor skills and your leg does not. This is due to the presence of different motor units.	
Evolution requires preexisting variation in the population, selective pressure, and reproduction. It happens at the population level.	
Many people have concerns about stem cells. Some of these are well-founded and others are not. You have to decide for yourself.	
Based on the data shown in this slide, researchers concluded that *Snarticus inferensis* is an invasive species.	
Think of a few examples of your own:	

Active Learning: Questions to Guide Discussion

1. What is active learning?
2. How do students construct new knowledge?
3. Some instructors claim that active learning takes too much time away from the content that needs to be covered. Do you agree or disagree? Explain.
4. True or false: Active learning should be fun and engaging.
5. Explain how active learning can foster a student-centered classroom.
6. Explain how active learning can be used to find out what students already know or what misconceptions they hold.
7. What teaching methods support active learning?
8. What is the role of motivation and emotion in learning?
9. Does active learning parallel the nature of science? Explain.
10. What motivates students to learn in your classroom?
11. Do you think that the teaching methods you use are effective? What evidence do you have?
12. Is active learning part of your personal teaching philosophy?
13. What are the learning goals when you teach? Do you choose activities that are designed to help students meet the goals?
14. Consider the terms in this list. How are they the same? How are they different? How do or could you use them in the classroom?

    Active learning
    Inquiry-based learning
    Student-centered learning
    Group learning
    Cooperative learning
    Constructivism

Scientific Teaching Workshop: Active Learning Resources

Key Concepts in Active Learning

Active learning engages students in thinking and behaving like scientists.

- Students learn when their current views (prior knowledge and misconceptions) are challenged by new information, and they have to construct new explanations to account for the information.
- Active learning provides a model of how scientists think and behave.
- Students who engage in challenging activities are more likely to take responsibility for learning.
- Group process and cooperation are integral components of active learning.
- Experimentation is the best way to engage students in active learning and it can easily be done in lecture or lab.

Suggested Reading

- Handelsman, J., S. Miller, C. Pfund. 2007. Chapter 2, Active Learning in *Scientific Teaching*. New York: W.H. Freeman & Co.
- DeHaan, R. L. 2005. The impending revolution in undergraduate science education. *Journal of Science Education and Technology* 14: 253-269.
- National Research Council 1999a. *How People Learn: Brain, Mind, Experience, and School.* Commission on Behavioral and Social Sciences and Education. Washington, D.C.: National Academies Press.
- National Research Council 1999b. *How People Learn: Bridging Research and Practice*. Commission on Behavioral and Social Sciences and Education. Washington, D.C.: National Academies Press.

Action Plan for Active Learning

Which tools, strategies, or resources from today's workshop will be most helpful to you in teaching this semester or in the near future?

Think of your last few lectures. What topics have you struggled to teach?

List two colleagues who could help you brainstorm active learning techniques to address the struggle.

1.

2.

Next time you teach, try one of the strategies that you and your colleague identify. How will you know if the strategy is successful? What will you observe or measure?

WORKSHOP III

Assessment

Workshop III: Assessment

Topic	Suggested Activities	Goals Participants will...
Welcome & Overview	Introduce goals and schedule of the workshop.	Know schedule and context for the workshop.
Assessment	1. Discuss case: "Frustrated Student" 2. Introduce assessment: Mini-lecture 3. Give examples: EnGauge-ments 4. Complete worksheet: "Compare and Contrast Two Assessments" 5. Complete worksheet: "EnGauge Students in Learning" 6. Discuss case: "Grading"	Engage in learning about assessment. Share prior knowledge, experience, and misconceptions about assessment Analyze knotty teaching problems and solve them collectively. Understand that regular, ongoing assessment is one of the most effective way to increase learning gains. Understand that assessment provides feedback to both students and instructors about learning; assessment techniques can be used to create an inclusive classroom, and assessment is more than grades. Explore assessment in depth and apply it to assessment in the classroom.
Summary	State key points of the workshop. Share resources. Do action plans.	Understand key points of the workshop. Have tools to apply assessment in the classroom. Reflect on the workshop.
Exit Assessment	Have participants write for one minute: *Explain how an assessment technique can simultaneously engage students in learning and help them gauge what they understand.*	Evaluate what was learned during this workshop.

Workshop Preparation

- Review Chapter 3.
- Review the cases, activities, and guiding questions.
- Make copies of the cases, resources, and action plan for participants.
- Make copies of the two worksheets.

Workshop III: Assessment

Notes and Tips for Workshop Facilitators

I. Goals

The goal of this workshop is to help participants understand that **assessment provides feedback to both instructors and students about learning, assessment drives student learning, it can be used to create an inclusive classroom, and it is more than grades**.

By the end of this workshop, participants will

- Understand that regular classroom assessment is one of the most effective ways to engage students in learning.
- Understand that assessment tools and active learning exercises can be combined; together, they simultaneously engage students in learning and gauge their learning.
- Understand that the results from assessments can be used to evaluate student learning, evaluate teaching, and guide changes in student behavior and instruction.
- Be able to solve knotty teaching problems with assessment techniques.
- Have developed a list of assessment techniques and resources that can be used in the classroom.
- Have developed an action plan for assessment.

II. Activities

1. **Discuss case: "Frustrated Student"**

   Participants read and discuss the "Frustrated Student" case.

2. **Introduce assessment: Mini-lecture**
   Use chapter 3 for ideas and information.
   The main points about assessment are that:
   a. Regular, ongoing assessment increases learning gains.
   b. Assessment should provide feedback to *both* students and instructors about learning.
   c. Assessment techniques can be used to create an inclusive classroom.
   d. Assessment is more than grades.
3. **Model examples of assessment techniques: EnGaugements**
   Model two or three "enGaugements" from chapter 3.
   Workshop participants should play the role of the students.
   After leading several enGaugements, discuss the process.
   a. What did the participants experience or learn?
   b. What feedback do enGaugements offer students? How does this compare with the feedback students get in a traditional classroom.
   c. How do enGaugements engage students in thinking and behaving like scientists?
   d. What are some benefits and drawbacks of using enGaugements?
4. **Complete worksheet: "Compare and Contrast Two Assessments"**
   Participants work with a partner to complete the worksheet. Discuss.
5. **Complete worksheet: "EnGauge Students in Learning"**
   Participants work in groups to complete the worksheet. Discuss.
6. **Discuss case: "Grading"**
   Participants read and discuss the "Grading" case.

Exit Assessment

Explain how an assessment technique can simultaneously engage students in learning and provide them with feedback about their own understanding.

Case: Frustrated Student

I am a junior majoring in biology. I was thinking I might go to graduate school to do research and become a professor, or maybe apply to medical school. I usually get A's in my courses; only a few B's so far in college. I totally breezed through high school; it was so easy.

This semester, I enrolled in introductory microbiology. I approach this class like most others: I attend lecture (have only missed two this semester!), read the textbook (usually before class, if I have time), and turn in the homework if it's going to be graded. Prof. Lopez is great; he's really well organized and follows the book closely. The homework has been helpful for learning the terms and information.

The first midterm exam in this course was NOT what I expected. None of the questions were multiple choice. We had to write out short (and sometimes LONG) answers. I barely finished it in the 2-hour exam period. Plus, three of the questions tested us on things we never learned and skipped stuff we had covered in class. For example, we learned about the *lac* operon last week, and it wasn't even on the test. But there was this question about asking us to "describe a strategy that bacteria use to regulate gene expression and explain why such a strategy might provide a selective advantage." How am I supposed to know about that? I got a 72% on that test. What a crock!

Forget microbiology; it's not for me.

Key questions to guide discussion:

- What issues might be contributing to this situation?
- Do the assessments give the students any feedback about what they understand while they are learning about this topic?
- What do the assessments motivate the student to learn? What effect do you think this professor's assessment will have on student behavior for the next test? Do you think that was the intention?
- What suggestions do you have for the professor?
- Have you faced a similar challenge?

Compare and Contrast Two Assessments

Work with a partner to compare and contrast what students experience during two different types of assessment activities.

	Case 1 Each week, students are assigned a reading. All students take a 10-min quiz that tests factual knowledge. Quizzes are handed in for points	**Case 2** Each week, students are assigned a reading. All students generate a diagram or flow-chart to illustrate the concept from the reading (individually). They explain their figure to each other in small groups for 10 minutes at the start of class. After discussion, they write a one-minute paper to explain what they learned. Diagrams and papers are handed in for points.
How does the assessment motivate students to learn the material or figure out the concepts they don't understand?		
How does the assessment capitalize on the diversity of learners?		
Does the assessment help students gauge what they know or how well they understand the concepts?		
Does the assessment build skills in collaboration and critical feedback?		
Write your own questions here:		

Assessment Worksheet: "EnGauge" Students in Learning

"EnGaugements" are activities that capture the spirit of scientific teaching: students simultaneously ***engage*** in learning and ***gauge*** what they are learning. Engaged students are more motivated to achieve the learning goals and take responsibility for learning, which is precisely the type of academic curiosity that effective science courses aim to awaken. A well-designed enGaugement motivates all students to learn and provides instructors and students with feedback about learning, and it integrates the three core themes of scientific teaching—active learning, assessment, and diversity. EnGaugements are particularly effective at addressing difficult concepts or skills, targeting common misconceptions, or emphasizing important points. Many enGaugements also lend themselves well to grading.

Instructions:

1. **Misconception:** List a common misconception in evolution (or another topic).
2. **Learning goal:** Write the correct version of the concept that students should understand.
3. **Intended learning outcomes:** Describe the specific performances or behaviors that will demonstrate whether students understand. (Use Bloom's Taxonomy or another resource to help articulate the outcomes.) In other words, what will it look like if students achieve the learning goal in the context of your classroom?
4. **EnGaugement:** Describe an activity that will "enGauge" students and help them achieve the learning goal.

Misconception	Learning goal	Intended learning outcomes	EnGaugement

Case: Grading

I attended a workshop about assessment, and the main thing I learned is that I am supposed to assess students before class so I can target what the students need to know. So, I created a series of pre-class quizzes for the students, but most students don't do them because they are not graded. However, I don't have the time to grade 320 of these each week—much less the 16 other assessments that the workshop suggested. I'll just go back to trusting my gut to know how well the students are doing.

Key questions to guide discussion:

- What issues might be contributing to this situation?
- What is this professor's definition of "assessment?"
- Other than grades, what strategies could motivate the students to participate in the assessments?
- What suggestions do you have for the professor?
- Have you faced a similar challenge?

Assessment: Questions to Guide Discussion

1. Explain how an assessment tool can simultaneously engage students in learning and gauge their learning.
2. Explain how feedback from assessment tools can guide changes in student behaviors.
3. True or false: assessments should always be graded by the instructor or the students won't know what they did wrong.
4. How can assessment foster a student-centered classroom? What opportunities will students have to gauge their progress toward the learning goals *during the learning process?*
5. What assessment tools do you use in your classroom? Do you think that they are effective? What evidence do you have?
6. How do you measure whether students are progressing toward the learning goals? How do you know whether they achieve the learning goals?
7. What evidence do you have that students do (or don't) understand the topic the way it's been taught in the past?
8. What evidence would convince you that an assessment tool is effective? What is "effective?"
9. Is assessment part of your personal teaching philosophy?
10. What are your goals when you teach? Do you choose assessment tools that are designed to provide you with information about your teaching and instruction?
11. Consider the terms in this list. How do they relate to each other? How do or could you use them in the classroom?

    Assessment tool
    Teaching method
    Evaluation
    Grading
    EnGaugement
    Active learning exercise
    Diversity

Scientific Teaching Workshop: Assessment Resources

Key Concepts in Assessment

Assessment drives student learning, it can be used to create an inclusive classroom, it is more than grades, and it provides feedback to both instructors and students about learning.

- ▶ Regular classroom assessment is one of the most effective ways to engage students in learning.
- ▶ Assessment tools and active learning exercises can be the same thing; together, they simultaneously engage students in learning and gauge their learning.
- ▶ Results from assessments can be used to evaluate student learning, evaluate teaching, and guide changes in student behavior and instruction.

Suggested Reading

- Handelsman, J., S. Miller, C. Pfund. 2007. Chapter 3, Assessment in *Scientific Teaching*. New York: W.H. Freeman & Co.
- Angelo, T. A., and K. P. Cross. 1993. *Classroom Assessment Techniques: A Handbook for College Teachers*. San Francisco: Jossey-Bass.
- Huba, M. E., and J. E. Freed. 2000. *Learner-Centered Assessment on College Campuses: Shifting the Focus from Teaching to Learning*. Needham Heights, MA: Allyn & Bacon.
- National Institute for Science Education, C. L. O. C.-T. 1999. *Field-Tested Learning Assessment Guide* (FLAG). Madison, WI: Wisconsin Center for Educational Research. http://www.flaguide.org
- National Institute for Science Education, C. L. O. C.-T. 1999. *Student Assessment of Learning Gains* (SALG). Madison, WI: Wisconsin Center for Educational Research. http://www.wcer.wisc.edu/salgains/instructor/
- Wiggins, G., and J. McTighe. 1998. *Understanding by Design*. Alexandria, VA: Association for Supervision and Curriculum Development.

Action Plan for Assessment

Which tools, strategies, or resources from today's workshop will be most helpful to you in teaching this semester or in the near future?

Think of your last few lectures. What topics have you struggled to teach?

List two colleagues who could help you brainstorm ways to address this struggle using assessment tools.

1.

2.

Next time you teach, try one of the strategies that you and your colleague identify. How will you know if the strategy is successful? What will you observe or measure?

WORKSHOP IV

Diversity

Workshop IV: Diversity

Topic	**Suggested Activities**	**Goals** Participants will...
Welcome & Overview	Introduce goals and schedule of the workshop.	Know schedule and context for the workshop.
Diversity	1. Discuss cases: "Cousins Vang", "Inclusive Classrooms", and "Bantering" 2. Introduce diversity: Mini-lecture 3. Think-pair-share: Attrition statistics 4. Model examples of diversity in the classroom 5. Complete worksheet: "Creating an Inclusive Classroom"	Engage in learning about diversity. Share prior knowledge, experience, and misconceptions about diversity. Analyze knotty teaching problems and solve them collectively. Understand that diversity of cognitive styles, teaching methods, and biases and assumptions affect learning. Explore diversity in depth and apply it to teaching.
Summary	State key points of the workshop. Share "Resources" page.	Understand key points of the workshop. Have tools to apply diversity in practice. Reflect on the workshop.
Exit Assessment	Write for one minute: *How does diversity affect learning?*	Evaluate what was learned during this workshop.

Workshop Preparation

- Review Chapter 4.
- Review the cases, activities, and guiding questions.
- Make copies of the cases, resources, and action plan for participants.

Workshop IV: Diversity

Notes and Tips for Workshop Facilitators

I. Goals

The goal of this workshop is to help participants understand that **diversity affects learning**; that knowledge, experience, and cultural influences make each student unique; that the teaching methods, examples, and content we choose to use affect who is included or excluded from the classroom experience; and that we all bring biases and assumptions to teaching and learning.

By the end of this workshop, participants will

- Understand that diversity is essential for progress in science.
- Understand that every teaching situation is unique and every student is unique.
- Understand that learning is idiosyncratic and culturally mediated (and therefore unique for every person, and different at different times).
- Understand that every person brings biases and assumptions to the classroom.
- Understand that a variety of teaching methods can reach a diversity of students.
- Be able to define "diversity" with regard to teaching methods, learning, and students.
- Be able to create opportunities in which diverse students learn from each other and enhance learning of the entire group.
- Be able to solve knotty teaching problems using a diversity of teaching methods.
- Have developed a list of tools, strategies, and resources that can be used in the classroom.

II. Activities

1. **Discuss case(s): "Cousins Vang", "Inclusive Classrooms", or "Bantering"**
   Participants read and discuss the cases.
2. **Introduce diversity: Mini-lecture**
   Use chapter 4 for ideas and information.

3. **Lead a "Think-pair-share" activity**
   Instructions for participants: Survey the statistics about attrition of students from science and engineering majors in college (Table 4.1 in this book). Propose a hypothesis to explain the attrition of students in the biological and physical sciences and engineering. Discuss with a partner.
4. **Model examples of ways to address diversity in science classrooms**
   Describe or teach two or three examples from chapter 4 in the section entitled, "Diversity in Practice" or from table 4.2. If appropriate, workshop participants should play the role of the students. Discuss.
5. **Complete worksheet: "Creating an Inclusive Classroom"**
   Participants work with a partner to identify strategies to make classroom experiences more inclusive.

III. Exit Assessment

How does diversity affect learning?

Case: Cousins Vang

I've taught introductory biology for three years now. Half of the grade is based on written work, such as lab reports and short-answer essay exams.

I've noticed that every semester I have two or three students whose last name is Vang. I asked one of the students if she was related to another Vang in the class. She said, "We're cousins." I also noticed that the Vang students seem to understand the material when I'm talking with them, yet their written answers tend to be unclear and often miss the point.

When I looked back at the grades for the past three years, I was surprised to find that no student with the last name Vang has ever received a grade higher than a C in my class. In fact, one-third of them have failed the course. I'm horrified! And I'm worried that my teaching style has some terrible, racist undertone that I never meant to perpetuate.

What can I do?

Key questions to guide discussion:

- What issues might be contributing to this situation?
- What suggestions do you have for the professor?
- Have you faced a similar challenge?
- Does it make you uncomfortable to have a last name listed in this case? What are the pros and cons to discussing a case with a name in it?
- What assumptions did you make about the professor and the students?

Case: Creating Inclusive Classrooms and Learning Experiences*

Following is an excerpt from a report that was written by the Learning through Evaluation, Adaptation, and Dissemination (LEAD) Center, entitled, *Minority Undergraduate Retention at UW-Madison: A Report on the Factors Influencing the Persistence of Today's Minority Undergraduates* (Alexander *et al.* 1998).

Excerpt from Report

Another factor was how UW's white undergraduates reacted to [minority student] presence. [I]t was the reaction of the white students that was most troubling to them and served to intensify their feelings of not belonging at the university.

> *It wasn't so much that the professors made the class hard, it was the study groups and stuff....Sometimes you're not even invited into the study groups. Or you ask them and they act uncomfortable. One day I was standing in the hallway and there was this group of white kids talking, and they basically said that the only way you come to this school as a black kid is if you're on an athletic scholarship of if they lowered the standards for you. And I was just like, whatever, I didn't even care. And then I thought to myself, "Okay, no matter how unaware you are, you should know better than to say that." ...And that's the kind of attitude you deal with a lot of times in this school. You don't even want to ask them to study with you. Or like, in my Spanish class last semester: the first day of class, I sit in the first seat closest to the door, and these white kids were breaking their necks to get as far away as they could. And the seat next to me did not fill up, the most convenient are right there next to me. It didn't fill until all the other seats filled. I mean, it kind of hurts my feelings, but it happens in all your classes, so you get used to it. I'm always guaranteed to have two seats next to me open, unless one of my black friends is in my class.*

The problem is not just that the overwhelming majority of undergraduates at UW-Madison are white, but that so many of them are racially uneducated whites from all-white communities. The interviewees who discussed prejudiced reactions from classmates generally believed the problem is more one of ignorance than of outright interracial hostility.

Key questions to guide discussion:

- What is your first reaction to reading this excerpt?
- If you witness these types of student-to-student interactions in your classroom, how might you, as an instructor, respond?
- Despite your best efforts to make a class as inclusive as possible, much of a student's learning occurs outside the classroom. As an instructor, what is your role in creating an inclusive learning environment outside the classroom?
- Can you relate to any of the student behaviors or reactions in the excerpt? What are they?

* This activity was adapted with permission from CIRTL (2006). The excerpt is from Alexander *et al.* (1998).

Case: Bantering

Professor:
"I use 'active learning' all the time in my class. Every minute or two, I ask questions. I think it makes me seem friendly and open to hearing the students' ideas. I try to call on students randomly to make sure everyone is engaged, but often I end up asking the bright students because they set the standard for the rest of the class."

Student #1:
"The professor is so engaging in this class. I feel like I have a connection with him, and I want to come prepared to class every lecture so I can answer the questions he asks. Sometimes, he asks a question, and I answer it, and then we have a great dialogue in front of the whole class about something way more advanced than what is in the syllabus. I am learning so much!"

Student #2:
"This class scares me to death. The professor asks questions every few minutes, and I am so afraid that he will ask me something I don't know the answer to. I have to sit toward the front because my hearing is poor, but I try to sit on the far right side of the room out of his line of sight so I can avoid eye contact. I can't wait until the semester is over so my anxiety can decrease."

Student #3:
"I have no clue what is going on in this class. The prof constantly asks questions, but I usually have no idea what the answer is. I must be really stupid because the students in the front seem to know all this already. Anyway, the prof ends up calling on them over and over. I either hide in the back so he won't call on me or skip lecture altogether. I wish he would just tell us what he wants us to know."

Key questions to guide discussion:

- What issues might be contributing to the differing opinions about the classroom experience?
- Who is included or excluded from this type of teaching?
- How does the professor define 'active learning'?
- Have you faced a similar challenge?
- What suggestions do you have for the professor?
- What assumptions or biases do you have about the professor or the students?

Diversity Worksheet: Creating an Inclusive Classroom

Work with a partner to identify strategies to make each situation more inclusive.

Teaching method or instructional choice	Who might feel excluded?	What could be done to make the classroom more inclusive?
Lectures are done with PowerPoint exclusively; the slides are dense with information. Notes are not available to students.		
Historical examples always involve white men.		
Examples always involve American people and situations.		
Exams are entirely multiple-choice and true-false.		
Exams are timed. Students are cut off after 2 hours. Grades are based on a curve.		
Homework assignments are only available online.		
The textbook costs $120.00.		
The class meets at 7:30 in the morning.		
Class requires students to work in assigned groups weekly, outside of class time.		

Diversity: Questions to Guide Discussion

1. What is "diversity"? Define it in the context of teaching and learning.
2. How does diversity affect student learning? In your classroom, does diversity enhance or impede learning?
3. What are your goals when you teach? Are they inclusive to all students? What evidence do you have?
4. How can you anticipate who may not be engaged by a particular teaching method?
5. In one or two sentences, describe your teaching philosophy. Does your teaching philosophy include any aspects of diversity?
6. Many science faculty claim they do not have enough time to address the individual needs of all their students. What arguments could you use to let them know that they don't need to become full-time advisors to reach diverse students?
7. Do you use examples or content that could be considered off-putting or offensive to students? How might this affect learning?
8. Explain the role of colleagues, support, and peer review in teaching for diversity.
9. What biases and assumptions do you bring to class? What can you do to minimize the impact of your biases on student learning?

Scientific Teaching Workshop: Diversity Resources

Key Concepts in Diversity

Diversity affects learning through the knowledge, experience, and cultural influences that make each student unique; through the teaching methods, examples, and content we choose to use; and through the biases and assumptions that everyone brings to the classroom.

- Diversity is essential for progress in science.
- Every teaching situation is unique and every student is unique.
- Learning is idiosyncratic and culturally mediated (and therefore unique for every person, and different at different times).
- Every person brings biases and assumptions to the classroom.
- A variety of teaching methods can reach more diverse students.

Suggested Reading

- Handelsman, J., S. Miller, C. Pfund. 2007. Chapter 4, Diversity in *Scientific Teaching*. New York: W.H. Freeman & Co.
- Women in Science and Engineering Leadership Institute. 2006. http://wiseli.engr.wisc.edu/
- Diversity Institute, Center for the Integration of Research, Teaching, and Learning. 2006. http://cirtl.wceruw.org/diversityinstitute/

Action Plan for Diversity

Which tools, strategies, or resources from today's workshop will be most helpful to you in teaching this semester or in the near future?

Think of your last few lectures. Which topics or teaching methods were not inclusive to all students?

List two colleagues who could help you brainstorm strategies to create a more inclusive classroom.

1.

2.

Next time you teach, try one of the strategies that you and your colleague identify. How will you know if the strategy is successful? What will you observe or measure?

WORKSHOP V

Institutional Transformation

Workshop V: Institutional Transformation

Topic	Suggested Activities	Goals Participants will...
Welcome & Overview	Introduce goals and schedule of the workshop.	Know schedule and context for the workshop.
Institutional transformation	1. Discuss cases: "Teaching Evaluations" and "Overextended" 2. Introduce institutional transformation: Mini-lecture 3. Lead a think-pair-share activity: Common misconceptions about teaching 4. Share examples and ideas for institutional transformation 5. Brainstorm and discuss strategies	Engage in learning about institutional transformation. Share prior knowledge, experience, and misconceptions about institutional transformation. Analyze challenges of institutional transformation and solve them collectively. Understand the importance of building a community of support in teaching and learning, leveraging administrative support and funding, and supporting recognition for teaching accomplishments. Explore institutional transformation in depth and instigate it in your department or on your campus.
Summary	State key points of the workshop. Share resources. Do action plans.	Understand key points of the workshop. Have tools to work on institutional transformation. Reflect on the workshop.
Exit Assessment	Have participants write for one minute: *What is your role in improving undergraduate science education on your campus?*	Evaluate what was learned during this workshop.

Workshop Preparation

- Review Chapter 6.
- Review the cases, activities, and guiding questions.
- Make copies of the cases, resources, and action plan for participants.

Workshop V: Institutional Transformation

Notes and Tips for Workshop Facilitators

I. Goals

The goal of this workshop is to help participants understand the importance of building a community of support in teaching and learning, leveraging administrative support and funding, and simultaneously giving and receiving recognition for teaching accomplishments in order to improve undergraduate science education.

By the end of this workshop, participants will

- Understand that institutional transformation can be started by one person or group.
- Understand that every campus and department is unique.
- Understand that instigating change requires building a network of colleagues and administrators who recognize and support teaching accomplishments.
- Be able to teach others how to solve knotty teaching problems using workshops and other strategies.
- Be able to use data from student learning to convince others that transformation is both necessary and feasible.
- Have developed a list of tools, strategies, and resources for transforming undergraduate science education on their campus or in their department.

II. Activities

1. **Discuss cases: "Teaching Evaluations" and "Overextended"**
   Participants read and discuss the cases.
2. **Introduce institutional transformation: Mini-lecture**
   Use chapter 6 for ideas and information.

3. **Lead a "Think-pair-share" activity**
   Instructions for participants: Survey the common misconceptions about teaching (Table 6.1 in this book). Discuss with a partner.
4. **Share examples of tools and ideas for institutional transformation**
   Give two or three examples of institutional transformation strategies from chapter 6. If appropriate, workshop participants should play relevant roles. Discuss. (Alternatively, have participants share their current struggles or questions, then collectively brainstorm solutions.)
5. **Brainstorm and discuss strategies**
   Brainstorm answers to the following questions. Discuss.
   a. **Change at the departmental level:**
      What departmental changes do you think are needed in teaching and learning? What evidence suggests they are needed? List three allies in your department who can support you in your efforts. How can you support them?
   b. **Change at the campus level:**
      What campus-wide changes do you think are needed in teaching and learning? What evidence suggests they are needed? List three administrators or campus programs that will be allies. How can they support your efforts?
   c. **Funding:** What funding opportunities are available to support these changes?

III. Exit Assessment

What would a "transformed" department look like? List three things you can do in science education in your department to move it toward achieving that vision.

Case: Teaching Evaluations

Professor Robinson, a faculty member in your department, agreed to teach the introductory course that no one wanted to teach. He tried several new teaching techniques in his course but got poor student evaluations. At the end of the semester, the curriculum committee asked the department chair to review his performance and take action.

Key questions to guide discussion:

- What issues might be contributing to this situation?
- What are the characteristics of an "effective" course? What role do student evaluations play in assessing the effectiveness of a course?
- What suggestions do you have for the professor?
- Are your responses different for pre-tenure vs. post-tenure faculty?
- What can you do, as a colleague?
- Have you faced a similar challenge?

Case: Overextended

Professor Dasgupta, a faculty member in your department, is an excellent teacher, as far as anyone can tell. She works tirelessly to make her classes engaging and challenging. Students do well in her class, on average, and they give her fantastic evaluations.

You are a member of her tenure committee and are concerned that she is spending too much time teaching, and even more on research. She is constantly pulled in several directions. How much longer can she keep up the pace? Something will have to give.

Key questions to guide discussion:

- What issues might be contributing to this situation?
- What suggestions do you have for the professor?
- What can you do, as a colleague and a member of the tenure committee?
- Have you faced a similar challenge?

Institutional Transformation: Questions to Guide Discussion

1. What are your goals in "institutional transformation?" What, exactly, do you hope to change? Why is this important to you? Why is it important for your department or campus?
2. In your department, what role do you play in supporting and recognizing your colleagues' teaching accomplishments? What is your role in training graduate students and postdocs to become faculty with scientific teaching skills? Can these be improved?
3. What colleagues, programs, and administrators are likely to support your efforts? What other departments might also help?
4. How can you anticipate who may be a barrier to change? How will you deal with those situations?
5. Change is a process. Explain the steps you should expect. How can you use these steps to show evidence of accomplishment?
6. In one or two sentences, describe your personal teaching philosophy. Does your teaching philosophy align with the university's mission? How could this affect the rate of change?
7. Many faculty in the sciences claim they do not have enough time to change how they teach. What arguments could you use to let them know why it's too important *not* to change how they teach? What could you say to convince them that they don't need to become full-time edcu-ational researchers to be scientific teachers?
8. Explain the role of colleagues, support, and peer review in institutional transformation.

Scientific Teaching Workshop: Institutional Transformation Resources

Key Concepts in Institutional Transformation

Change happens in stages; institutional change is no exception.

Important aspects of institutional transformation include:

- Building a community of support in teaching and learning.
- Leveraging administrative support and funding.
- Giving/receiving recognition for teaching accomplishments.
- Using a variety of approaches to target more people and magnify the effect.

Suggested Readings

- Handelsman, J., S. Miller, C. Pfund. 2007. Chapter 6, Institutional Transformation *in* Scientific Teaching. New York: W.H. Freeman & Co.
- Women in Science and Engineering Leadership Institute. 2006. http://wiseli.engr.wisc.edu/

Action Plan for Institutional Transformation

Which tools, strategies, or resources from today's workshop will be most helpful to you?

What issues have you struggled with in changing science education at your campus?

List two colleagues who could help you brainstorm what campus-wide or departmental changes could improve science education and the issues you point out above. With your colleagues, identify a few strategies for change, and enlist their help and support.

1.

2.

How will you know whether the strategy is successful? What will you observe or measure? Remember to use the five stages of change.

List possible sources of funding:

List possible campus resources, such as teaching and learning centers, office for faculty professional development, graduate TA training program, or a supportive provost:

References

Abrams, F. 2005 (May 20). Cognitive conundrum. *The Times Educational Supplement*, p.18.

Alexander, B.B., J. Foertsch, D. Bowcock, and S. Kosciuk. 1998. *Minority undergraduate retention at UW-Madison: A report on the factors influencing the persistence of today's minority undergraduates.* Madison, WI: University of Wisconsin-Madison LEAD Center.

Allen, D. and B. Duch. 1998. *Thinking toward solutions: Problem-based learning activities for general biology*. Philadelphia: Saunders College Publishing.

Angelo, T.A., ed. 1998. *Classroom assessment and research: An update on uses, approaches, and research findings*. San Francisco: Jossey-Bass.

Angelo, T.A. and K.P. Cross. 1993. *Classroom assessment techniques: A handbook for college teachers*. San Francisco: Jossey-Bass.

Angier, N. 1991 (April 4). Molecular 'hot spot' hints at a cause of liver cancer. *New York Times.*

Antonio, A. 2002. Faculty of color reconsidered: Reassessing contributions to scholarship. *Journal of Higher Education* 73:582-602.

Astin, A.W. 1993. Diversity and multiculturalism on the campus: How are students affected? *Change* 25:44-50.

Ausubel, D. 1963. *The psychology of meaningful verbal learning.* New York: Grune & Stratton.

___. 2000. *The acquisition and retention of knowledge: A cognitive view.* Boston: Kluwer Academic Publishers.

Beichner, R., L. Bernold, E. Burniston, P. Dail, R. Felder, J. Gastineau, M. Gjertsen, and J. Risley. 1999. Case study of the physics component of an integrated curriculum. *Physics Education Research* 67:S16-S24.

Beichner, R. and J.M. Saul. 2003. Introduction to the SCALE-UP (Student-Centered Activities for Large Enrollment Undergraduate Programs) project. *Proceedings of the International School of Physics.*

Biernat, M. and M. Manis. 1994. Shifting standards and stereotype-based judgments. *Journal of Personality and Social Psychology* 66:5-20.

Bieron, J. and F. Dinan. 1999. *Case studies across a science curriculum.* Department of Chemistry and Biochemistry, Buffalo, NY: Canisius College.

Black, P. and D. Wiliam. 1998. Inside the black box: Raising standards through classroom assessment. *Phi Delta Kappan* 80(2):139-148.

Blair, I.V., J.E. Ma, and A.P. Lenton. 2001. Imagining stereotypes away: The moderation of implicit stereotypes through mental imagery. *Journal of Personality and Social Psychology* 81:828-841.

Bloom, B.S. and D.R. Krathwohl. 1956. *Taxonomy of educational objectives: The classification of educational goals, by a committee of college and university examiners. Handbook 1: Cognitive domain.* New York: Longmans.

Bolea, B. 2006. Letter to the editor: An overemphasis on learning styles. *The Chronicle of Higher Education* 52:55.

Bransford, J.D., A.L. Brown, and R.R. Cocking. 2000. *How people learn: Brain, mind, experience, and school committee on developments in the science of learning*. Washington, DC: National Academy Press.

Brew, A. 2001. *Research-led teaching: What does it look like and how and why should academic developers encourage it?* Paper presented at the annual conference of the Staff and Educational Development Association, Manchester, UK, November 20-21.

Bruer, J.T. 2006. Points of view: On the implications of neuroscience research for science teaching and learning: Are there any? A skeptical theme and variations: The primacy of psychology in the science of learning. *CBE—Life Sciences Education* 5:104-110.

Bybee, R. 1993. An instructional model for science education. In *Developing biological literacy*. Colorado Springs, CO: Biological Sciences Curriculum.

Carnes, M., J. Handelsman, and J. Sheridan. 2005. Diversity in academic medicine: The stages of change model. *Journal of Women's Health* 14:471-475.

CIRTL. 2006. Center for the Integration of Research, Teaching, and Learning. http://cirtl.wceruw.org/ (accessed July 25, 2006).

Coffield, F., D. Moseley, E. Hall, and K. Ecclestone. 2004. *Learning styles and pedagogy in post-16 learning: A systematic and critical review*. London: Learning and Skills Research Centre.

Connolly, M.R., J.L. Bouwma-Gearhart, and M.A. Clifford. 2006. The birth of a notion: The windfalls and pitfalls of tailoring a SoTL-like concept to scientists, mathematicians, and engineers. *Innovative Higher Education*. In press.

Cox, J. 1993. *Cultural diversity in organizations: Theory, research and practice*. San Francisco: Berrett-Koehler Publishers.

Crosling, G., and G. Webb. 2002. *Supporting student learning: Case studies, experience and practice from higher education*. London: Kogan Page.

Cross, K.P. 1990. Teaching to improve learning. *Journal on Excellence in College Teaching* 1:9-22.

Cross, K.P. and T.A. Angelo. 1988. *Classroom assessment techniques: A handbook for faculty*. Ann Arbor: University of Michigan, National Center for Research to Improve Postsecondary Teaching and Learning.

Cross, K.P. and M.H. Steadman. 1996. *Classroom research: Implementing the scholarship of teaching*. San Francisco: Jossey-Bass.

Culver, R.S. and J.T. Hackos. 1982. Perry's model of intellectual development. *Engineering Education* 72: 221-226.

Curry, L. 1983. An organization of learning styles theory and construct. *Educational Research Information Center (ERIC) Document No. ED 235 185*.

___. 1987. *Integrating concepts of cognitive or learning style: A review with attention to psychometric standards*. Ottawa, Ontario: Canadian College of Health Service Executives.

Dasgupta, N. and A.G. Greenwald. 2001. On the malleability of automatic attitudes: Combating automatic prejudice with images of admired and disliked individuals. *Journal of Personality and Social Psychology* 81:800-814.

Davies, P.G., S.J. Spencer, D.M. Quinn, and R. Gerhardstein. 2002. Consuming images: How television commercials that elicit stereotype threat can restrain women academically and professionally. *Personality and Social Psychology Bulletin* 28(12):1615-1628.

Deaux, K. and T. Emswiller. 1974. Explanations of successful performance on sex-linked tasks: What is skill for the male is luck for the female. *Journal of Personality and Social Psychology* 29:80-85.

Dees, R.L. 1991. The role of cooperative learning in increasing problem-solving ability in a college remedial course. *Journal for Research in Mathematics Education* 22:409-421.

DeHaan, R.L. 2005. The impending revolution in undergraduate science education. *Journal of Science Education and Technology* 14:253-269.

Desmedt, E., M. Valcke, L. Carrette, and A. Derese. 2003. Comparing the learning styles of medicine and pedagogical sciences students. In *Bridging theory and practice.* S. Armstrong, M. Graff, C. Lashley, E. Peterson, S. Raynor, E. Sadler-Smith, M. Schiering, and D. Spicer, eds. Proceedings of the Eighth Annual European Learning Styles Information Network Conference, University of Hull.

Deutsch, M. 1949a. An experimental study of the effects of cooperation and competition upon group process. *Human Relations* 2:199-231.

___. 1949b. A theory of cooperation and competition. *Human Relations* 2:129-152.

Dewey, J. 1916. *Democracy and education.* New York: MacMillan, The Free Press.

Duch, B., S. Groh, and D. Allen, eds. 2001. *The power of problem-based learning: A practical 'how to' for teaching undergraduate courses in any discipline.* Sterling, VA: Stylus Publishing, LLC.

Dunn, R. and K. Dunn. 1992. *Teaching secondary students through their individual learning styles.* Needham Heights, MA: Allyn and Bacon.

Duren, P.E. and A. Cherrington. 1992. The effects of cooperative group work versus independent practice on the learning of some problem-solving strategies. *School Science and Mathematics* 92:80-83.

Ebert-May, D. and E. P. Weber. 2006. FIRST—What's next? *CBE—Life Sciences Education* 5:27-28.

Ebert-May, D., J. Batzli, and H. Lim. 2003. Disciplinary research strategies for assessment of learning. *BioScience* 53:1221-1228.

Edwards, H., B. Smith, and G. Webb, eds. 2001. *Lecturing: Case studies, experience and practice.* London: Kogan Page.

Eisenkraft, A. 2003. Expanding the 5E model. *The Science Teacher* 70(6):56-59.

Entwistle, N, ed. 1990. Teaching and the quality of learning in higher education. In *Handbook of educational ideas and practice.* London: Routledge.

Evans, M.E., H. Schweingruber, and H.W. Stevenson. 2002. Gender differences in interest and knowledge acquisition: The United States, Taiwan, and Japan. *Sex Roles* 47:153-167.

Fagen, A.P., C.H. Crouch, and E. Mazur. 2002. Peer instruction: Results from a range of classrooms. *Physics Teacher* 40:206-209.

Felder, R.M., J.E. Stice, and A. Rugarcia. 2000. The future of engineering education VI: Making reform happen. *Chemical Engineering Education* 34:208-215.

Felder, R.M., D.R. Woods, J.E. Stice, and A. Rugarcia. 2000. The future of engineering education II: Teaching methods that work. *Chemical Engineering Education* 34:26-39.

Gardner, H. 1983. *Frames of mind: The theory of multiple intelligences.* New York: Basic Books.

Garner, I. 2000. Problems and inconsistencies with Kolb's learning styles. *Educational Psychology* 20(3):341-348.

Glassick, C.E., M.T. Huber, and G.I. Maeroff. 1997. *Scholarship assessed: Evaluation of the professoriate.* San Francisco: Jossey-Bass.

Goldin, C. and C. Rouse. 2000. Orchestrating impartiality: The impact of blind auditions on female musicians. *American Economic Review* 90:715-741.

Grimmett, P.P. and G.L. Erickson. 1988. *Reflection in teacher education.* New York: Teachers College Press.

Guimerà, R., B. Uzzi, J. Spiro, and L.A. Nunes Amaral. 2005. Team assembly mechanisms determine collaboration network structure and team performance. *Science* 308:697-702.

Gurin, P. 1999. Expert Report of Patricia Gurin, *Gratz,* et al. *v. Bollinger,* et al., No. 97-75321 (E.D. Mich.) & *Grutter,* et al. *v. Bollinger,* et al., No. 97-75928 (E.D. Mich.). *The Compelling Need for Diversity in Higher Education.*

___. 2002. Diversity and higher education: Theory and impact on educational outcomes. *Harvard Educational Review* 72:330-366.

Hake, R.R. 2002. Lessons from the physics education reform effort. *Conservation Ecology* 5(2):28-67.

Halme, D.G., J. Khodor, R. Mitchell, and G.C. Walker. 2006. A small-scale concept-based laboratory component: The best of both worlds. *CBE—Life Sciences Education* 5:41-51.

Handelsman, J., D. Ebert-May, R. Beichner, P. Bruns, A. Chang, R. DeHaan, J. Gentile, S. (Miller) Lauffer, J. Stewart, S.M. Tilghman, and W.B. Wood. 2004. Scientific teaching. *Science* 304:521-522.

Hastings, S. 2005 (November 4). Learning styles. *The Times Educational Supplement,* p.11.

Hatch, J., M. Jensen, and R. Moore. 2005. Manna from heaven or clickers from hell. *Journal of College Science Teaching* 34:36-39.

Hattie, J. 1999. *Influences on student learning.* Inaugural lecture, University of Auckland, New Zealand.

Haynes, N.M. and S. Gebreyesus. 1992. Cooperative learning: A case for African-American students. *School Psychology Review* 21:577-585.

Heilman, M.E. and M.H. Stopeck. 1985. Attractiveness and corporate success: Different causal attributions for males and females. *Journal of Applied Psychology* 70:379-388.

Hershey, D.R. 2004. *Avoid misconceptions when teaching about plants.* Actionbioscience.org. http://www.actionbioscience.org/education/hershey.html (accessed May 25, 2006).

Hogsett, C. 1992. Women's ways of knowing Bloom's taxonomy. *Feminist Teacher* 7(3): 27-32.

Honey, P. and A. Mumford. 1982. *Manual of learning styles.* London: Peter Honey.

Huba, M.E. and J.E. Freed. 2000. *Learner-centered assessment on college campuses: Shifting the focus from teaching to learning.* Needham Heights, MA: Allyn & Bacon.

Huber, M.T. 1999. *Developing discourse communities around the scholarship of teaching.* Paper presented at the annual meeting of the American Association of Higher Education, Washington, DC.

Huber, M.T. and P. Hutchings. 2005. *The advancement of learning: Building the teaching commons.* San Francisco: Jossey-Bass Publishers.

Hutchings, P., C. Bjork, and M. Babb. 2002. The scholarship of teaching and learning in higher education: An annotated bibliography. *Political Science* 35:233-236.

Hutchings, P. and L.S. Shulman. 1999. The scholarship of teaching: New elaborations, new developments. *Change* 31:10-15.

Hyde, J.S. 2005. The gender similarities hypothesis. *American Psychologist* 60:581-592.

Hyde, J.S., E. Fennema, and S. Lamon. 1990. Gender differences in mathematics performance: A meta-analysis. *Psychological Bulletin* 107:139-155.

Hyde, J.S. and M.C. Linn. 1988. Gender differences in verbal ability: A meta-analysis. *Psychological Bulletin* 104:53-69.

Inzlicht, M. and T. Ben-Zeev. 2000. A threatening intellectual environment: Why females are susceptible to experiencing problem-solving deficits in the presence of males. *Psychological Science* 11(5):365-371.

Isaak, M. 2003. *Five major misconceptions about evolution.* The Talk Origins Archive. http://www.talkorigins.org/faqs/faq-misconceptions.html (accessed May 25, 2006).

Jagers, R.J. 1992. Attitudes toward academic interdependence and learning outcomes in two learning contexts. *Journal of Negro Education* 61:531-538.

Johnson, D.W. and R.T. Johnson. 1975. *Learning together and alone: Cooperation, competition, and individualization*. Englewood Cliffs, NJ: Prentice Hall.

Johnson, D.W., R.T. Johnson, and L. Scott. 1978. The effects of cooperative and individualized instruction on student attitudes and achievement. *Journal of Social Psychology* 104:207-216.

___. 1985. Classroom conflict: Controversy versus debate in learning groups. *American Education Research Journal* 22:237-256.

Kember, D. 2000. *Action learning and action research: Improving the quality of teaching and learning.* London: Kogan Page.

Khodor, J., D.G. Halme, and G.C. Walker. 2004. A hierarchical biology concept framework: A tool for course design. *Cell Biology Education* 3:111-121.

Kieser, J., P. Herbison, and T. Harland. 2005. The influence of context on students' approaches to learning: A case study. *European Journal of Dental Education* 9:150-156.

Knight, J.K. and W.B. Wood. 2005. Teaching more by lecturing less. *Cell Biology Education* 4:298-310.

Kolodny, A. 1991. Colleges must recognize students' cognitive styles and cultural backgrounds. *The Chronicle of Higher Education* 37:A44.

Kratzig, G.P. and K.D. Arbuthnott. 2006. Perceptual learning style and learned proficiency: A test of the hypothesis. *Journal of Educational Psychology* 98:238-246.

Kulis, S., Y. Chong, and H. Shaw. 1999. Discriminatory organizational contexts and black scientists on postsecondary faculties. *Research in Higher Education* 40(2):115-148.

Labov, J.B. 2003. From the National Academies: Overview of two ongoing activities at the National Academies of interest to CBE readers. *Cell Biology Education* 2:202-204.

___. 2005. From the National Academies: Ongoing challenges to evolution education: Resources and activities of the National Academies. *Cell Biology Education* 4:269-272.

Lawrence, P.A. 2006. Men, women, and ghosts in science. *Public Library of Science (PLoS) Biology* 4(1):e19.

Lawson, A.E. 2006. Points of view: On the implications of neuroscience research for science teaching and learning: Are there any? *CBE—Life Sciences Education* 5:111-117.

Lawson, A.E. and M. Johnson. 2002. The validity of Kolb learning styles and neo-Piagetian developmental levels in college biology. *Studies in Higher Education* 27:79-90.

Lazarowitz, R., R.L. Hertz, J.H. Baird, and V. Bowlden. 1988. Academic achievement and on-task behavior of high school biology students instructed in a cooperative small investigative group. *Science Education* 72:67-71.

Little Soldier, L. 1989. Cooperative learning and the Native American student. *Phi Delta Kappan* 71:161-163.

Loo, R. 2004. Kolb's learning styles and learning preferences: Is there a linkage? *Educational Psychology* 24:99-108

Lujan, H.L. and S.E. DiCarlo. 2006. First-year medical students prefer multiple learning styles. *Advances in Physiology Education* 30:13-16.

Lummis, M. and H.W. Stevenson. 1990. Gender differences in beliefs and achievement: A cross-cultural study. *Developmental Psychology* 26(2):254-263.

Macbeth, D. 2000. On an actual apparatus for conceptual change. *Science Education* 84:228-264.

Marzano, R.J. 1998. *A theory-based meta-analysis of research on instruction.* Aurora, CO: Mid-continental Regional Educational Laboratory (McREL).

Mazur, E. 1996. *Peer interaction, a user's manual.* Upper Saddle River, NJ: Prentice Hall.

McIntosh, P. 1990. White privilege: Unpacking the invisible knapsack. *Independent School* 49(2):31-36.

McLeod, P.O., S.A. Lobel, and T.H. Cox. 1996. Ethnic diversity and creativity in small groups. *Small Group Research* 27:248-265.

McManus, D. 2005. *Leaving the lectern: Cooperative learning and the critical first days of students working in groups.* Bolton, MA: Anker Publishing.

Mervis, J. 2006. NIH told to get serious about giving minorities a hand. *Science* 311:328-329.

Mettetal, G. 2001. The what, why, and how of classroom action research. *The Journal of Scholarship of Teaching and Learning* 2(1):6-13.

Milem, J.F. 2001. Increasing diversity benefits: How campus climate and teaching methods affect student outcomes. In *Diversity challenged: Evidences on the impact of affirmative action.* G. Orfield, ed.. Cambridge, MA: Harvard Education Publishing Group.

Modell, H., J. Michael, and M.P. Wenderoth. 2005. Helping the learner to learn: The role of uncovering misconceptions. *The American Biology Teacher* 67:20-26.

National Center for Education Statistics. 2000. *Digest of Educational Statistics.* Washington, DC: Department of Education.

National Institute for Science Education. 1999a. *Field-tested Learning Assessment Guide (FLAG).* Wisconsin Center for Educational Research. http://www.flaguide.org (accessed August 23, 2006).

___. 1999b. *Student Assessment of Learning Gains (SALG).* Wisconsin Center for Educational Research. http://www.wcer.wisc.edu/salgains/instructor/ (accessed August 23, 2006).

National Research Council. 1999a. *How people learn: Brain, mind experience, and school.* Commission on Behavioral and Social Sciences and Education, Washington, DC: National Academies Press.

___. 1999b. *How people learn: Bridging research and practice.* Commission on Behavioral and Social Sciences and Education. Washington, DC: National Academy Press.

___. 1996. *National Science Education Standards.* Washington, DC: National Academy Press.

___. 2003. *Bio2010: Transforming undergraduate education for future research biologists. Committee on Undergraduate Biology Education to Prepare Research Scientists for the 21st Century,* Board on Life Sciences, Division on Earth and Life Studies, Washington, DC: National Academy Press.

___. 2005. *Assessment of NIH minority rResearch training programs, phase 3.* Committee for the Assessment of NIH Minority Research Training Programs, Board on Higher Education and Workforce. Washington, DC: National Academies Press.

___. 2006. *Rising above the gathering storm: Energizing and employing America for a brighter economic future.* Committee on Prospering in the Global Economy of the 21st Century: An Agenda for American Science and Technology. Washington, DC: National Academies Press.

National Science Board. 2002. *Science and Engineering Indicators.* Arlington, Virginia: National Science Foundation.

Nelson, D.J. 2004. Nelson diversity surveys. *Diversity in Science.* http://cheminfo.chem.ou.edu/~djn/diversity/top50.html (accessed October 23, 2006).

Nelson, S., and G. Pellet. 1997. *Shattering the silences* [video recording]. New York: Gail Pellet Productions.

Nemeth, C.J. 1985. Dissent, group process, and creativity: The contribution of minority influence. *Advances in Group Processes* 2:57-75.

___. 1995. Dissent as driving cognition, attitudes, and judgments. *Social Cognition* 13:273-291.

Novak, J.D. and A.J. Canas. 2006. The theory underlying concept maps and how to construct them. *Technical Report IHMC CmapTools* 2006-01. Pensacola, FL: Florida Institute for Human and Machine Cognition.

Okebukola, P.A. 1986a. Cooperative learning and students' attitudes to laboratory work. *School Science and Mathematics* 86:582-590.

___. 1986b. The influence of preferred learning styles on cooperative learning in science. *Science Education* 70:509-517.

Olian, J.D., D.P. Schwab, and Y. Haberfeld. 1988. The impact of applicant gender compared to qualifications on hiring recommendations: A meta-analysis of experimental studies. *Organizational Behavior and Human Decision Processes* 41:180-195.

Pfundt, H. and R. Duit. 1994. *Bibliography: Students' alternative frameworks and science education.* Kiel, Germany: University of Kiel Institute for Science Education.

Plato. 1901. *The Republic.* New York: Colonial Press.

Pribbenow, C.M., J. Sheridan, M. Carnes, E. Fine, and J. Handelsman. 2006. The department chair and climate: Contradicting perceptions. In progress.

Prochaska, J.O. and C.C. DiClemente. 1983. Stages and processes of self-change of smoking: Toward an integrative model of change. *Journal of Consulting and Clinical Psychology* 51:390-395.

Prochaska, J.M., J.O. Prochaska, and D.A. Levesque. 2001. A transtheoretical approach to changing organizations. *Administration and Policy in Mental Health* 28:247-261.

Quinn, D.M. and S.J. Spencer. 2001. The interference of stereotype threat with women's generation of mathematical problem-solving strategies. *Journal of Social Issues* 57(1):55-71.

Reiff, J.C. 1992. *Learning Styles.* Washington, DC: National Education Association.

Rewey, K.L., D.F. Dansereau, S.M. Dees, L.P. Skaggs, and U. Pitre. 1992. Scripted cooperation and knowledge map supplements: Effects on the recall of biological and statistical information. *The Journal of Experimental Education* 93-107.

Reynolds, M. 1997. Learning styles: A critique. *Management Learning* 28(2):115-133.

Rosser, S. 1990. *Female-friendly science*. New York: Teachers College Press.

Rutherford, F.J. 1990. *Science for all Americans*. New York: Oxford University Press.

Sagaria, M.A.D. 2002. An exploratory model of filtering in administrative searches: Toward counter-hegemonic discourses. *The Journal of Higher Education* 73(6):677-710.

Schneps, M.H. 1989. *A private universe* [video recording]. San Francisco, CA: Astronomical Society of the Pacific.

Seymour, E. and N.M. Hewitt. 1997. *Talking about leaving: Why undergraduates leave the sciences*. Boulder, CO: Westview Press.

Shih, M., T.L. Pittinsky, and N. Ambady. 1999. Stereotype susceptibility: Identity salience and shifts in quantitative performance. *Psychological Science* 10(1):80-83.

Siu, H.M., H.K. Spence Laschinger, and E. Vingilis. 2005. The effect of problem-based learning on nursing students' perceptions of empowerment. *Journal of Nursing Education* 44:459-470.

Shulman, L.S. 1999. Taking learning seriously. *Change* 31:11-17.

Smith, M.E., C.C. Hinckley, and G.L. Volk. 1991. Cooperative learning in the undergraduate laboratory. *Journal of Chemical Education* 68:413-415.

Sommers, S.R. 2006. On racial diversity and group decision making: Identifying multiple effects of racial composition on jury deliberations. *Journal of Personality and Social Psychology* 90(4):597-612.

Spelke, E.S. 2005. Sex differences in intrinsic aptitude for mathematics and science? A critical review. *American Psychologist* 60(9):950-958.

Spencer, S.J., C.M. Steele, and D.M. Quinn. 1999. Stereotype threat and women's math performance. *Journal of Experimental and Social Psychology* 35:4-28.

Spring, L., M.E. Stanne, and S. Donovan. 1999. Undergraduates in science, mathematics, engineering, and technology: A meta-analysis. *Review of Educational Research* 69:21-51.

Steinpreis, R.E., K.A. Anders, and D. Ritzke. 1999. The impact of gender on the review of the curricula vitae of job applicants and tenure candidates: A national empirical study. *Sex Roles* 41:509-527.

Sun Microsystems. 2006. *Global inclusion: Dimensions of diversity*. http://www.sun.com/aboutsun/globalinclusion/dimensions.html (accessed June 4, 2006).

Swisher, K. 1990. Cooperative learning and the education of American Indian/Alaskan native students: A review of the literature and suggestions for implementation. *Journal of American Indian Education* 29:36-43.

Tanner, K. and D. Allen. 2006. Approaches to biology teaching and learning: On integrating pedagogical training into the graduate experiences of future science faculty. *CBE—Life Sciences Education* 5:1-6.

Tobias, S. 1990. *They're not dumb, they're different: Stalking the second tier*. Tucson, AZ: Research Corporation.

Turner, C.S.V. 2000. New faces, new knowledge. *Academe* 36:34-37.

Udovic, D., D. Morris, A. Dickman, J. Postlethwait, and P. Wetherwax. 2002. Workshop biology: Demonstrating the effectiveness of active learning in an introductory biology course. *BioScience* 52:272-281.

Uhlmann, E.L. and G.L. Cohen. 2005. Constructed criteria: redefining merit to justify discrimination. *Psychological Science* 16:474-480.

U.S. Census. 2000. U.S. Department of Commerce, Bureau of the Census. *Census of population and housing, 2000.* Washington, DC: United States Summary Files.

Vermunt, J.D. 1996. Metacognitive, cognitive and affective aspects of learning styles and strategies: A phenomenographic analysis. *Higher Education* 31:25-50.

Walters. M.R. 1999. Case-stimulated learning within endocrine physiology lectures: An approach applicable to other disciplines. *Advances in Physiology Education* 276:74-78.

Waterman, M.A. and E.D. Stanley. 2005. *Biological Inquiry: A workbook of investigative case studies.* San Francisco: Benjamin Cummings.

Weiss, E.M., G. Kemmler, E.A. Deisenhammer, W. W. Fleischhacker, and M. Delazer. 2003. Sex differences in cognitive functions. *Personality and Individual Differences* 35:863-875.

Wells, D.G., M. McEvoy, and M. Kundel. 2006. Review: Learning the hard way. *CBE—Life Sciences Education* 5:123-125.

Wenneras, C., and A. Wold. 1997. Nepotism and sexism in peer-review. *Nature* 387:341.

Wiggins, G. and J. McTighe. 1998. *Understanding by design.* Alexandria, VA: Association for Supervision and Curriculum Development.

Wilkerson, L. and W.H. Gijselaers, eds. 1996. *Bringing problem-based learning to higher education.* New Directions for Teaching and Learning Series, No. 68. San Francisco: Jossey-Bass.

Wood, W.B. and J. Gentile. 2003. Meeting report: The first National Academies Summer Institute for undergraduate education in biology. *Cell Biology Education* 2:207-209.

Wood, W.B. 2004. Clickers: A teaching gimmick that works. *Developmental Cell* 7:796-798.

Wood, W.B. and J. Handelsman. 2004. Meeting report: The 2004 National Academies Summer Institute on undergraduate education in biology. *Cell Biology Education* 3:215-217.

Woods, D.R. 1994. *Problem-based learning: How to gain the most from PBL.* Hamilton, Ontario: D. R. Woods.

___. 1995. *Problem-based learning: helping your students gain the most from PBL.* Waterdown, Ontario: D.R. Woods.

Wulff, D.H., ed. 2005. *Aligning for learning: Strategies for teaching effectiveness.* Bolton, MA: Anker Publishing Company.

Zardetto-Smith, A.M. 2006. Review: A new formula for better learning: A cup of common sense and a dash of neuroscience? *CBE—Life Sciences Education* 5:126-127.

Index